Motivation – The Gender Perspective of Young People's Images of Science, Engineering and Technology (SET)

Felizitas Sagebiel (ed.)

Motivation – The Gender Perspective of Young People's Images of Science, Engineering and Technology (SET)

Proceedings of the Final Conference

Budrich UniPress Ltd.
Opladen • Berlin • Toronto 2013

Bibliografische Information der Deutschen Nationalbibliothek
Die Deutsche Nationalbibliothek verzeichnet diese Publikation in der Deutschen Nationalbibliografie; detaillierte bibliografische Daten sind im Internet über http://dnb.d-nb.de abrufbar.

Gedruckt auf säurefreiem und alterungsbeständigem Papier.

Alle Rechte vorbehalten.
© 2013 Budrich UniPress, Opladen, Berlin & Toronto
www.budrich-unipress.de

ISBN 978-3-940755-81-0 (Paperback)

Das Werk einschließlich aller seiner Teile ist urheberrechtlich geschützt. Jede Verwertung außerhalb der engen Grenzen des Urheberrechtsgesetzes ist ohne Zustimmung des Verlages unzulässig und strafbar. Das gilt insbesondere für Vervielfältigungen, Übersetzungen, Mikroverfilmungen und die Einspeicherung und Verarbeitung in elektronischen Systemen.

Typografisches Lektorat: Judith Henning, Hamburg – www. buchfinken.com
Umschlaggestaltung: disegno visuelle kommunikation, Wuppertal – www.disenjo.de
Druck: paper&tinta, Warschau
Printed in Europe

Table of contents

MOTIVATION. Gender, image and choice of science and
engineering. An introduction (Felizitas Sagebiel) 7

Socialization agents and the gendered choice of educational
paths: Perpetuation or fragility of gender stereotypes?
(Susana Vázquez-Cupeiro) 29

Part 1 – Images of SET, youth and gender 45

Why girls stay away from STEM: How the image of science
clashes with teenagers' identity (Ursula Kessels) 47

Images of an ideal engineer and self-image – differences between
male and female engineering and non-engineering students
(Martina Endepohls-Ulpe and Judith Ebach) 61

Uncertainty as a barrier in job decision making process
of young women (Kathrin Gräßle) 75

The gender perspective of young people's images of
science, engineering and technology (SET) in the Slovak
Republic (Nataša Urbančíková and Gabriela Koľveková) 83

Secondary school students' reactions to descriptions of
engineering and nursing in university catalogues (Minna
Salminen-Karlsson) 99

Young people and nanotechnologies (Ilse Marschalek,
Petra Moser and Magdalena Strasser) 115

Part 2 – Peers, school and media 129

Does science education at school make young Europeans
like SET? (Anne-Sophie Godfroy) 131

Teach the teachers: Gender competence as an innovative
element of teacher-training in mathematics (Anina Mischau,
Bettina Langfeldt and Karin Griffiths) 145

ICT resources in STEM activities and ICT and STEM
career prospects for young students: Gender-related
issues (Àgueda Gras-Velázquez, Alexa Joyce and
Albert Gras-Martí) 159

Part 3 – Methodology and evaluation of good practice 171

Are young people lazy, blind, or misguided?
A fresh look at unpleasant facts (Frank Stefan Becker) 173

Mind the gap: Science and young people (Federica Manzoli,
Flora Di Martino, Daniele Gouthier, Donato Ramani) 201

Munich's gender-sensitive education – extend girls' world
to the fascinating field of engineering Motivating girls to go
for jobs with interesting tasks, career prospects and steady
incomes (Barbara Roth) 219

Mentoring and video clips – ways against stereotype
threat (Sylvia Neuhäuser-Metternich and Sybille Krummacher) 239

Inspiring girls and boys in science, engineering and
technology on virtual networks – the e-mentoring
programme TANDEMkids at RWTH Aachen University
(Marcel Lämmerhirt and Carmen Leicht-Scholten) 253

Mentoring programme TANDEMschool – encouraging
young people to study SET subjects by a coherent MINT
cooperation concept at RWTH Aachen University (Gehrt
Hartjen and Carmen Leicht-Scholten) 271

How to change stereotypical images of science, engineering &
technology? Results and conclusions from the european project
MOTIVATION (Felizitas Sagebiel (Germany), Carme Alemany
(Spain), Jennifer Dahmen (Germany), Bulle Davidsson (Sweden),
Anne-Sophie Godfroy-Genin (France), Gabriela Kol'veková (Slovakia),
Cloé Pinault (France), Els Rommes (The Netherlands), Mariska
Schönberger (The Netherlands), Anita Thaler (Austria), Natasa
Urbancíková (Slowakia), Christine Wächter (Austria)) 287

Authors' short biographies 293

MOTIVATION. Gender, image and choice of science and engineering. An introduction

Felizitas Sagebiel

This introduction has two functions, first to give basic information about the background and methodology of the MOTIVATION project[1], and second to give an overview about the structure and content of the book.

1 Introduction to the MOTIVATION project

The European Commission project MOTIVATION[2] – "Promoting positive images of SET in young people under gender perspective" as coordination action within the 7th Framework Programme aimed on exchanging information between the partner countries about individual and societal factors, which influence the image of science, engineering and technology (SET) and its consequences for gendered study and occupational choices. Participants were universities and non-profit associations from seven countries (Austria, France, Germany, Netherlands, Slovakia, Spain, Sweden).

Questions for this study came from results from previous European projects[3] in which most of the partners had also cooperated. In order to attract

---

1 This volume is gathering papers hold on the final conference of MOTIVATION project under the heading of "The Gender Perspective of Young People's Images of Science, Engineering and Technology (SET)" which took place from 10th to 12th December 2009 in the Department of Educational and Social Sciences at the University of Wuppertal.
2 This introduction will give basic methodological information of MOTIVATION project while a summary of the results will be given at the end of this volume where a short article about the project will included (Sagebiel et al. 2009) and be reprinted.
3 Other European projects on similar issues at the University of Wuppertal have been: INDECS (www.indeces.uni-wuppertal.de) (Partner countries: Austria, Finland, France, Germany, Greece, Slovakia, Spain, UK), coordination: Professor Dr Clivia Sotomayor-Torres, University of Wuppertal, which proved the attractiveness and acceptability of inter-disciplinary degree courses on data basis of a one-year explanatory research (2001-2002); Womeng "Creating Cultures of success for women engineers" (www.womeng.net) (Partner countries: Austria, Finland, France, Germany, Greece, Slovakia, Spain, UK), coordination: Conférence des Directeurs d'Ecoles Françaises d'Ingénieurs (CDFI) by Dr. Yvonne Pourrat and Ecole Normale Supérieure de Cachan (Dr. Anne-Sophie Godroy-Genin, scientific coordination), which investigated engineering and non-engineering students and asked them about images, knowledge and attitudes of engineering. At the same time organisational cultures of engineering departments were studied by interviewing of male and female faculty (2002-2005), and PROMETEA "Empowering Women Engineers in Industrial and Academic Research" (www.prometea.info) 13 partner countries (Austria, Finland, France, Germany, Greece, Lithuania, Russia, Serbia, Slovakia, Spain, Sweden, UK, and Chile) with core partners: Austria, Finland, France, Germany, Slovakia (dissemination), Spain, UK, coordina-

more girls and women for science, engineering and technology (SET), as well as similar interested boys, in the first instance several problematic elements of degree courses have to be changed. Research from European Project Womeng (Sagebiel 2007; Thaler/Wächter 2005) as well as from the former European Project INDECS (Sagebiel 2005; Sagebiel 2007) indicate that interdisciplinarity in engineering education (Wächter 2005) is needed to attract students, particularly female students, to engineering. Interviews in companies showed that technology organisations would appreciate getting more interdisciplinary trained engineers. Interdisciplinarity as well as single sex education in engineering degree courses, it is argued, could help to decrease gender stereotypes (Sagebiel 2007: 157). In particular, single sex teaching could increase self-confidence in young women. However this proposal is not welcomed by faculty members in higher education – including female members – because they fear that this "unrealistic" educational setting would not prepare female engineers for their professional life. From interviewing faculty in engineering degree courses, it has emerged that a traditional masculine image of engineering is still alive and evidenced in competitive male interactions and sexist language and humour (Sagebiel 2007: 159; Sagebiel/Dahmen 2006). Women, it was found, react with different coping strategies depending on how equal the culture surrounding them is characterised; in cases of subtle discrimination the female student tends to laugh them off, while the female student can be left feeling isolated in cases of more blatant discrimination (Sagebiel 2007: 159). The traditional engineering culture has meant that an interdisciplinary approach has been neglected, as well as a project and team oriented pedagogy (Sagebiel 2007: 158), and there has been no change to the traditional image of degree courses nor the male oriented culture of departments. Moreover, research showed that all students, even engineering students, complained about the lack of job information about engineering professions (Dahmen 2006). And even though many recruitment measures try to transmit an image of optimal job opportunities in the future, homepages for engineering degree courses and departments, investigated in European project Womeng, were generally found not to reflect this new image (Sagebiel/Dahmen 2006).[4]

With this background, the MOTIVATION project developed the idea of having a greater understanding of the interconnecting reasons why young people do not choose SET very often. The project took gender and migration background into account, focussing the process on issues that pre-existed the

---

tion: Conférence des Directeurs d'Ecoles Françaises d'Ingénieurs (CDFI) by Dr. Yvonne Pourrat and Ecole Normale Supérieure de Cachan (Dr. Anne-Sophie Godroy-Genin, scientific coordination), which investigated career possibilities, promoters and barriers for women's engineering research (2005-2007).

4   The proceedings of the final conference of MOTIVATION will show the summarised results as well as information about similar projects on the issue. For a short German version of the conference results also see Sagebiel/Dahmen (2010b).

decision for or against studying SET. Often young people have obsolete and unattractive job images about SET in their minds, when, in fact, their hoped-for professions are not far off from a SET professional reality. So, one thought was that with better information students could become more interested in SET. Just as socialisation agents, peer groups, teachers and media give information and are influential, so the attitudes of young people towards SET jobs differ according to gender (Sagebiel/Dahmen 2010a). Factors could include how young people were informed and which role models were presented and how they were perceived by them. The use of realistic role models in professional engineering could lead to making the relevant jobs more attractive by changing the image of these professions and in this way increasing the number of interested students, especially female students.

2   State of the art of the MOTIVATION project

In the MOTIVATION project there are several theoretical terms and concepts (causes and effects) relevant to the issues of gendered choice of SET training and studying that are interconnected with each other: Gender stereotypes; gender awareness; socialisation agents (like media, teacher, peers, study advisors); science education; teacher's education, knowledge about SET; professional associations; image of SET; pictures of SET and professions; perception of SET and professions; study and professional choices; professional associations; and young people's gender pressure.

Gender stereotypes are a central reason why girls (in Western Europe) less often choose for science and engineering subjects, at school and in further education. SET is gendered, which means that they are traditionally male associated and constructed (see Wajcman 1996; Faulkner 2000). Wajcman blames dichotomized thinking here: "[m]ales are portrayed as fascinated with the machine itself, 'being' hard masters'…Females are described as only interested in computers as tools" (Wajcman 1996: 56). Such images reproduce gender stereotypes of men and women" (Sagebiel 2007: 151). According to Faulkner (2000), engineering is constructed as gendered in three aspects: The gendered labour division as different working styles, the symbolic connections between technique and dominant masculinity, and the gendered connotation of professional identities. There is a stereotype that SET is better suited to boys than girls. This broad societal stereotype has to be changed when it comes to science education in school. Teachers have to learn about gender stereotypes and gender awareness in their own education to avoid reproduction of gender stereotypical choice of subjects of their pupils. Schools are constructing gender differences through the attitudes of teachers, in addition to the influence of peers, gendered learning material, and didactics (Duru Bellat 2004; Zaidman 1992).

To create knowledge about SET professions in a modern way is a strategy that has to be developed by professional associations as well as institutions of higher education and professional persons as role models. This knowledge has to be spread in society as a whole and especially to the socialisation agents in order to change images and attitudes. Young people should also be a major focus, so that the masculine image of SET changes to a more gender neutral image and perception of SET and its associated professions become less biased. Chances for less gendered study and professional choices could then be increased. Looking to the differential role of socialisation agents, it is obvious that media works most effectively via images. Thus, it can be said that relevant media like magazines and television series or soaps, working as today's informal 'vocational counsellors' for young people (Dostal & Troll 2002; Jim Studie 2006), can also influence teenagers' images of SET. Many teenagers, even after finishing school, do not know which study or vocational training they should choose. At this point, the knowledge and attitudes of career and study advisors can become very important. However, by way of example, Beinke's (2006) study showed that the people in these roles are often not qualified to professionally advise young people.

Gender stereotypes not only focus on SET's image and knowledge of it, in general they become very strong during puberty so that teenagers of both gender try to adapt to traditional gender expectations. According to peer group expectations, girls behave carefully to be considered feminine and boys try to be seen as "real boys". As decisions for or against SET are made step by step in puberty, both sides of gender stereotypes, that is, the professional image of SET and the gender pressure for the young people, have to be changed to ensure the possibility of gender balanced choices of study and training subjects. The connection between gender images in puberty and gendered study choice has been investigated and theorized by Hannover and Kessels[5] (for instance Hannover & Kessels 2004). Gendered identity formation as the most crucial factor for study and job decision for young people has also been studied by other authors (e.g. Henwood 1996; Rasmussen 1997; Faulkner 2000; Corneliussen 2002; Rommes et al. 2007; Schreiner and Sjøberg 2006). Interest in the subjects is a necessary condition for subject and professional choice. The Rose project showed that, "Out-of-school experiences in physics" is strongly correlated to "interest in physics" (ROSE project, Trumper 2006), indicating that beyond the school environment there are other possibilities for influence. Experiences with measures taken with girls in education show that special single-sex recruitment methods for SET directed at girls to interest and support them in studying these subjects (like the Ada-Lovelace Project, www.ada-mentoring.de 26.08.06) can be effective.

---

[5] A paper by Kessels on this issue has been included in this book.

State of the art as the first step in the MOTIVATION project (see Sagebiel, Dahmen 2010a) showed a different amount of data regarding the impact of socialisation agents on young people's job attitudes. While for instance in Germany there seems to be a lack of research on how SET is presented in youth relevant TV series, in soap operas (Götz 2003) or in movies (Steinke 2003) – as well as how people's jobs are generally presented on TV –, school socialisation has been analysed more often, such as like teachers (Dimbath 2007) and the role of peer groups (Beinke 2004). Job orientation processes of girls and young women were investigated by Nissen et al. (Nissen et al. 2003). A large amount of pedagogical research exists in Germany on the issue of how more girls could be attracted for natural sciences at school and on how the SET lessons could be improved and updated (e.g. Faulstich-Wieland 2004). How job titles can influence young people's perception was studied by Krewerth et al. (2004) with a research pool of 900 male and female pupils. Choice of vocational training or degree courses is dependent on the image of a job in society as well as whether it has a modern job title. However, the dream jobs of boys and girls remain gender stereotypical (Sagebiel/Dahmen 2010a).

While the amount and focus of national studies varied from partner country to partner country the international ROSE-study filled the gaps between these variations. In this study 40.000 pupils in 35 countries were asked about their perceptions of and attitudes towards science and technology (Schreiner/Sjöberg 2006). The results show a gender stereotypical perception of jobs that also influence the attitudes of teenagers towards SET.

Many national initiatives or websites aiming on sensitizing teachers for gender topics in general exist, but it's not clear how many persons are reached by those initiatives and because of a lack of evaluation the impact can be questioned. The annual "Girls' Day" in Germany seems successful, being directly evaluated afterwards with a questionnaire (Wentzel 2005), but the evaluation of the long-term impact of the participation on girls is lacking here too. So it is not clear if the participation of girls will influence job and study choices.

3   Methodology

To answer the question as to how gender stereotypes were reproduced, the MOTIVATION project focussed on the following objectives (Sagebiel et al. 2008): Exchange about what has been done in research, evaluation of content, methods and didactics of information about SET under gender aspects, and understanding interdependencies with gendered job decisions, collecting measures of good practice, evaluating them and creating new effective methods for changing images of SET under gender aspects.

Starting from general project objectives, research questions were developed based on the state of the art and structured in different work packages (wp). These work packages were structured in defined objectives, broken down into tasks and instruments to fulfil the tasks. Deliverables as products they had also to be defined focusing on different factors for making job choices: One work package focused on the influence of print and visual media (wp2)[6], a second on teachers and advisors influence (wp3), a third work package investigated young people's images of self in connection to job decisions (wp4), and a fourth work package collected so called good practices of influencing teenagers job interests in each partner country (wp5).

The different work packages (wp) were coordinated by the different individuals in the project team, Anita Thaler, Anne-Sophie Godfroy-Genin, Els Rommes, Bulle Davidsson, and Natasa Urbancikova. Each work package started with hypotheses for which special methods were designed on collective basis. A draft was designed by the work package leader and afterwards it was completed by the other team members of the project.

For work package 2, "Youth, gender and SET in media", coordinated for the Austrian partner by Anita Thaler, three hypotheses were developed: 1) Movies, television series, music clips and magazines contribute in constructing culturally dominant images; 2) Experiences through media are influential for learning of interests; and 3) Relevant media such as magazines and television series are today's informal vocational counsellors for young people. To study these hypotheses content analyses of youth magazines and soaps in TV programmes about how gender stereotypes in science and technology are represented were done. Relevant mass media for teenagers in the age bracket 15 to 16, such as the most important national youth magazines and TV soaps, were systematically analysed by using a developed data sheet to know how gender and stereotypes in science and technology are represented.

For Work package 3, "Youth, gender and SET at school and beyond", coordinated by Anne-Sophie Godfroy-Genin, three hypotheses were developed: 1) School, unconsciously, creates exclusion through images, habits, language, learning material, teachers' attitudes; 2) Gender differences are built at school through different ways, through peer relations among the children, through attitudes of teachers towards children, often unconsciously gendered, and through learning material, representing gendered images; and 3) There is rare mention of women in history or in history of sciences and examples used in mathematical problem solving or in sciences present slight gender discrimination.

Based on state of the art and a comparative overview of different school systems several methodological measures were constructed, practised and evaluated in the field work. Different secondary school types with different

---

6 This work package will not be an issue of this paper.

student populations by class and ethnicity were chosen for interviews and focus group discussions to get their perspectives on SET individually and in relation to their peers as well as expert interviews with teachers about their perspectives and about their perceived influence on the attitudes of teenagers. Complementarily, an exemplary documentary analysis of school books in use was also carried out.

For work package 4, "Consequences of gendered public and self-images for SET job decisions", coordinated by Els Rommes, three hypotheses were created: 1) Gendered identity formation might be the most crucial factor for study choice; 2) Adolescents tend to choose their study based on a prototype of someone working in a profession, even when they know this prototype is incorrect, and even when this prototype includes characteristics that are irrelevant for that profession; and 3) Adolescents systematically compare what they are good at, what they want from a job, and what activities they like, with their (in)correct expectations of a particular profession that is a prototype of someone working in a profession.

Even though a division of labour existed in work packages in terms of conceptualising and realising methodological steps, issues from different work packages were included in one methodological instrument. Interviews and focus group discussions with young people about their images of SET, occupational choices, and possible peer group influence were combined for work package 3 and 4. Common guidelines for interviews with pupils and teachers were developed by the persons responsible for wp3 and wp4, and then exchanged and coordinated between other team project members by email. In an expert workshop a discussion was held with experts focussing on migrant aspects in school. Guidelines for focus group discussions with pupils were developed in the same way.

Content, methods and didactics of information about SET under gender aspects were analysed for understanding interdependencies with gendered job decisions. In gender awareness raising workshops the attitudes of teachers as well as study and career consultants were exchanged.

For work package 5, "Good Practice of changing measures for images of SET for both genders", coordinated by Bulle Davidsson, the following four objectives were postulated: 1) To evaluate good practice in media presentation about images of SET; 2) to exchange measures for changing image of SET in school and beyond taking gender in account; 3) to exchange measures for raising gender and SET awareness in consultants (advisors) for girls and boys in job decision situations; and 4) to exchange good practice with relevant stakeholders.

Good practice projects from the Netherlands were presented by experts and compared with national initiatives of the other partner countries. For each partner country two examples of good practice were evaluated by interviews with the persons responsible. On this basis, qualitative case studies were described, and results of case studies and evaluations lead to developing and designing a website as the recommended main action (Sagebiel et al. 2008).

A specialty in this coordination action has been the so-called expert meetings with invited external scientists for knowledge exchange. These persons included Minna Salminen-Karlsson, who gave a paper on "good practice" and also held discussions with the partners and helped to further develop the next steps in methodological approach. The second expert workshop worked in similar ways, where topics for all work packages were included. Expert input was given for the gendered image of SET in print media, the method of content analysis in print media and participants' experiences with this topic and method were discussed. The same method was practiced for the work package focussing on the influence of gendered SET images in school on young people, expert presentation about the topic, discussion of methods, and knowledge exchange between participants. For the good practice experiences, expert presentation was completed by international comparison.

4   Theoretical perspectives of gender, image and choice of science and engineering

The MOTIVATION project covered three strands of thinking and understanding as to why science and engineering is less attractive for girls and why there is a broader interest among boys. The socialisation factors, teachers, peer groups and media, and best practice were considered in terms of how to change the situation.

To give the following articles a broader scope we also refer to the European literature review of "Meta-analysis of gender and science research" (European Commission 2012) because the issues should not be considered in terms of single factors but, rather, the greater environment of society with its associated genderedness has also to be taken in account. It has been found that (horizontal) gender segregation in science starts with gender segregation (Meulders et al. 2010a) in education for which socialisation and gender stereotypes (Sagebiel/Vázquez Cupeiro 2010) are very important.

There is a complexity dilemma for changing the situation and changing (reducing) gender stereotypes. Concrete initiatives can only grip some factors and it is not possible to manipulate all and especially not the more complex ones. So, even if a change measure could be theoretically effective the effect can be destroyed if the society is, for instance, less gender-balanced. While societal influences work on the macro level, others, such as school systems, can be effective on choice of science subjects on a meso level. To cite Caprile et al.: "Recent studies show that structural and life-course factors play a major role, both in terms of educational achievement and study field choices.

Differences are less pronounced in integrated educational systems and in more gender equal societies" (European Commission 2012: 72).

Moreover, taking the social construction of women and girls into account there is always an essentialism trap when looking for more female orientation in media presentations, curriculum, and pedagogy in school as well as in higher education, and in the diverse recruitment initiatives. And, by referring to essential effects, gender stereotypes, which should be reduced for a gender balanced choice of science subjects, can be reinforced again.

A bulk of research studies has been done on gender differences regarding cognitive abilities focussing on maths and science, "…in most cases the underlying assumption is that girl's under-achievement in maths is the main cause of the under-representation of women in scientific and technological studies, a fact which is not confirmed by recent studies" (Alauf et al. 2003a; Xie and Shauman 2003; after European Commission 2012: 73). Ordinarily, these results are exaggerated and there seems to be a special interest in not giving up this research tradition (see Cordelia Fine's criticism[7] 2012). The so constructed differences are made responsible for the gender segregation in science. Nevertheless, the regular PISA studies of OECD seem to refer to this strong research tradition.

However, as research has shown repeatedly, abilities too are socially constructed by environmental influences as well as by the individuals themselves. So, parents seem to under-evaluate the science abilities of their daughters just as the daughters under-evaluate their capabilities (Abele 2002; European Commission 2012

Girls seem to perceive social acceptance in their process of developing their identity as more important. Thus, if people in their environment think that SET is not appropriate for girls, girls will try to adapt to these expectations. As there is an association between technical abilities and interests with masculinity in society, this will have a negative effect on girls' choice of SET. Girls will react to this in their feminine identity construction, especially in puberty when the gender stereotypes are reinforced.

The following overview of the papers combines preponderantly papers being presented at the final conference of the MOTIVATION project under the heading of "The Gender Perspective of Young People's Images of Science, Engineering and Technology (SET)".

---

7   At the European conference on «Gender and Higher Education» 2012 at Bergen (Norway) she gave a brilliant key note speech on her criticism of this research stream (see also Fine 2010).

5 Overview of structure and content

Susana Vázquez-Cupeiro starts with her overview of relevant research literature in the article on "Socialization agents and the gendered choice of educational paths: Perpetuation or fragility of gender stereotypes?"[8] From her view, gender roles are socially constructed and vary according to the culture and over the course of people's lives, particularly from childhood into adolescence. As key agents of socialization she refers to family (mainly parents), school (teachers), peers and mass media (and increasingly ICT) which play a key part in the transmission of gender roles. The literature review shows that in school the negative influence of teachers and academic advisors, the male-biased technology curriculum, together with the small number of women teachers who serve as role models to younger students are the main arguments used to explain the segregation of women in SET. Youth culture is not favourable to technical subjects (seen as more difficult) and this perception may be decisive when it comes to choosing the field of study, first at school and later at university. During their leisure time children and teenagers are also influenced by gender-related stereotypes that may predispose them to make certain career choices and not others. The media and ICT are powerful agents for transmitting gender roles and pervasive stereotypes. As a result, while there is little doubt that the media present largely traditional gender images, the evidence of the impact of such images on gender attitudes and behaviour is not only inconclusive but it can also be used to promote a gender-bias-free socialization. A considerable part of the literature emphasizes the need to transform the contexts of learning in order to make them more inclusive of girls (European Commission 2012).

The structure of the following section mostly follows the conference structure. The articles are structured in three parts, "Images of SET, youth and gender", "Peers, school and media", and "Methodology and evaluation of good practice". The three parts are not identical with the work packages of the project, they are overlapping with different work packages and they consist of more issues than those combined in one work package. Each part will begin with a leading paper that acts as an overview, followed by more specialised papers in these thematic areas[9]. Not all conference papers were cho-

---

[8] This paper uses data from the thematic report written by Felizitas Sagebiel and Susana Vázquez-Cupeiro "Stereotypes and Identity" which is part of the European Commission funded tender *Meta-analysis of gender and science research* (2008-2010), under the 7th RTD European Framework Programme commissioned by the DG Research, led by CIREM (Spain).

[9] At this point I want to thank social scientist Jennifer Dahmen who worked as team member for the University of Wuppertal and organized the conference under the responsibility of associate professor Dr Felizitas Sagebiel as project leader. She cooperated in preparing part of this publication. The following information about the papers is mostly based on abstracts of the authors.

sen and some new papers that fitted with the issue of the volume were included[10].

5.1 Images of SET, youth and gender

During adolescence young people often struggle with their self-perception. Personal changes in combination with the trials of fulfilling societal expectations of being either "feminine" or "masculine" can influence individuals' perception of gender-role "appropriate" jobs. In their direct environment, in- and outside of school, teenagers face different values and norms that are of great importance during this developmental phase. This topic deals with individuals' interconnection of SET images and gender and discusses which attitudes male and female teenagers have towards SET. How does gender influence the interest in these fields? And generally, what aspirations do young girls and boys have regarding their future?

As leading paper, Ursula Kessels' article on "Why girls stay away from STEM: How the image of science clashes with teenagers' identity" has been included because its theory and research is very well developed and founded. Concerning female students' low involvement in science related subjects she emphasizes the role that the formation of *academic interests* has for adolescents' *identity development in general*. Based on several studies regarding the *perceived fit/misfit between adolescents' self-image and the image of science* she shows that the typical girl that likes and excels in physics was perceived as being very unfeminine, and considered masculine. Additionally, other aspects of the typical representative of science seem highly incompatible with the self-image of most (female) adolescents.

Students disapprove of science to the extent that their self-image deviates from the image of the typical student favouring science. If students perceive themselves as being very different from the typical student favouring science, they usually report not to like science, not to aim at a science related career and they do not choose a science related profile at school.

The stereotyping of physical science as a masculine domain makes stronger engagement in physics threatening for female adolescents, because it endangers their developing identity as a "woman-to-be", while avoiding physics will foster and stabilize their gender-identity. Presented will be also intervention studies aimed at enhancing the perceived fit between girls' self-image and the image of science related subjects.

Martina Endepohls-Ulpe and Judith Ebach titled their paper "Images of an ideal engineer and self-image – differences between male and female engineering and non-engineering students". Their study is part of the Europe-

---

10 After the conference the authors took the opportunity to revise their papers. Most of papers have been reviewed several times.

an research project UPDATE (Understanding and Providing a Developmental Approach to Technology Education), which aims at improving science and technology teaching in Europe, especially for girls. They examined why girls drop out of technology education at different stages of their educational career.

Based on a questionnaire study, consisting of 179 non-engineering students and 141 engineering students[11], results have revealed that the image of an ideal engineer is in general closely related to the male stereotype. The authors believe that the differences with respect to the self-images of male and female engineering and non-engineering students reflect these stereotypes. However, the self-image of female engineering students appears to be closer to their own image of an ideal engineer than those of their female peers in non-technological studies. Furthermore, their images of an ideal engineer seem to be the most modern or progressive of all subgroups.

Kathrin Gräßle in her German study investigates "Emotions and images – barriers preventing young women from choosing technical courses"[12]. Her research questions were what the causes are that lead to young women studying engineering subjects less often than men. She found two main reasons, which interact and reinforce each other. The first reason is the emotion „uncertainty". While choosing a course to study female pupils want to get rid of this uncertainty as soon as possible. There are different strategies of decision making, not all of which lead to a technical option. As the consideration of a technical course is regarded as abnormal, the young women tend to disregard it. This corresponds with the second reason: The incompatible images between being a woman, and their images of technology careers. Gräßle's methodology is both qualitative and longitudinal and based on interviews with ten female pupils. In order to get more female students universities should adapt their didactics on orientation methods: Offering support, giving young women the chance to reflect on their situation and creating personal networks – in short: Making emotions and images comprehensible.

Natasa Urbancikova and Gabriela Kol'veková present Slovakian results from the Motivation project in "The gender perspective of young people's images of science, engineering and technology (SET) in Slovak Republic". First on the state of the art within Slovak education system they point out the socialistic footprints in the rebirth of democracy in the new state. In their results they particularly found the so-called habits of "gift" and "credit". The results show that the peer group influence prevails over the teachers influence on SET images and job decisions. Moreover, media are a strong influential factor on all young people. The TV soap opera analysis proved that television is reflecting some consequences of gendered public images for SET deci-

---

11 The methodological measure was influenced by the European Womeng project (Pourrat 2005).
12 Gräßle's article is based on the study for her published thesis (Gräßle 2009).

sions. Teachers were exactly aware of the dream/nightmare jobs of their pupils.

The next study is a Swedish one. Minna Salminen-Karlsson asked "How do adolescents react to formulations about engineering and nursing in university catalogues?" The paper reports on a questionnaire study on how Swedish secondary school students reacted to university recruitment material about study programmes in engineering and hospital nursing.

Nursing programmes seem to prepare their students to being a member in a collective. Engineering programmes presented themselves as "almost like having a job", downplaying the academic content, seeing the students as autonomous and often giving them a lot of choices. They seemed to prepare their students into finding an individual niche in the job market. Based on these results 366 secondary school students answered a questionnaire where they were asked to choose one of the four constructed programmes and comment on them, as well as answer a number of questions about their preferences when choosing university education. The results show that when the students could not associate the descriptions to certain professions, the choices were relatively gender balanced. The results point at a number of aspects that could be changed in the organization of studies or just the descriptions in the information material to make engineering programmes more attractive to female applicants.

Ilse Marschalek, Petra Moser and Magdalena Strasser report in their paper on "Young people and nanotechnologies" on the EU seventh framework programme funded project NANOYOU. Their work package on initial user requirements included a comprehensive qualitative and quantitative survey to identify young people's knowledge, attitudes, specific values, concerns and expectations concerning nanotechnologies in general, and in the three sub areas, that is, information and communication technologies, energy and environment, and medicine and health in more detail. The survey considered young people between the ages of 11 and 25. These age parameters were used because, firstly, they are seen as a critical public who accepts and adopts new technologies, and will be the consumers of new products, and secondly, because they are seen as the future engineers and scientists in the various sectors and fields of nanotechnologies. The survey wanted to especially focus on finding out the different female and male ways of thinking and their interests and how to find out particular ways to attract females and their needs for communication. The core methical approaches were: Focus group discussions with young people in Austria, the UK, and Israel; expert interviews with experts in different fields concerned, as well as interviews with science centres' representatives who are part of the NANOYOU consortium; an online-survey in eight languages and three age groups, and national context survey with representatives of each participating project country.

5.2 Peers, school and media

Besides their parents, peers, school, and media[13] can be seen as main influencing instances for young people in occupational or study decision making processes. Although the awareness about the impact of these socialization agents exists, for instance in youth relevant media meaningful and gender equal SET representations are still rare to find. It will be discussed how the popularity and image of SET is co-shaped by each of the three instances and in what ways and to what extent they are interconnected. What influence do they have for young people in decision making processes for or against SET? Finally, the role of science education in weakening or strengthening a positive image of SET is discussed.

As part of the MOTIVATION project, Anne-Sophie Godfroy-Genin's paper on "Does science education at school make young Europeans like SET?" focuses on images of science communicated through education at school. In particular she studied how a gendered image of SET is constructed at schools in the different European partner countries and whether the images created are attractive or not. Different patterns of science education were compared to the percentage of boys and girls choosing SET in higher education. Based on research by Marie Duru Bellat (2004 and before), interviews were done with pupils aged between 15 and 18 and teachers. Complementary discussion focus groups were also an instrument to survey attitudes of the teenagers in the same age groups. In each country two different schools were chosen as case studies, one in a rather privileged area, one in a rather underprivileged area. In addition, document analysis of school books in SET and general information about school systems in different partner countries were carried out and exchanged. Some attempts are presented to improve the situation through various measures at school, including: New teaching methods, specific teaching material, single-sex classes in SET, interdisciplinary and problem-based approaches.

"Teach the teachers: Gender competence as an innovative element of teacher-training in mathematics" by Anina Mischau, Bettina Langfeldt and Karin Grabarz is based on the project "Gender competence as an innovative element within the professionalization of teacher-training in mathematics" funded by the Ministry for Education and Research. National as well as international comparative studies show that mathematics must still be seen as a "typical boys' subject" (cf. Baumert et al. 1997; Bos et al. 2008; Zimmer 2004). The authors problematize this because, besides its own relevance as a school subject or a scientific discipline, mathematics is also a "core discipline" for the field of natural sciences and engineering (cf. DMV 2007). De-

---

[13] Media as socialization agent will not be focused on in a special article, but results are integrated in Sagebiel et al. (2009), which summarized the results of the MOTIVATION project and will be reprinted here at the end of the volume.

velopment and implementation of gender sensitive didactics as well as the necessity of a sensitisation of maths teachers for their contribution to creating and reproducing gender stereotyped domains seems necessary (cf. Ziegler et al. 1998; Keller 1998; Buchmayer 2008). Gender stereotyping and the lack of gender competence on the part of teachers must be identified as fundamental reasons for the subject-specific gender bias on the development of interests and competences of young people.

The project tackles two deficiencies: The lack of practice-orientated education and the lack of a gender perspective within the mediation of subject specific and didactic contents (cf. Mischau et al. 2009).

Àgueda Gras-Velázquez, Alexa Joyce and Albert Gras-Marti present their research on "How to encourage students and teachers to look forward to MST classes: Five projects, five visions". They looked at projects to motivate young people, especially girls and at teachers' education. For instance the Stella handbook on "Innovative practices in science education for European schools" targeting educational authorities, school heads and science teachers, analyses the main factors that bring and diffuse innovation in science teaching at schools. It is illustrated with the pool of good practices identified during the project life span, and providing a summary of the most relevant content and debates featured in the web portal. ITEMS provided off-the-shelf Moodle courses on different Science and Mathematics topics and workshops to train teachers in their use.

5.3 Methodology and Evaluation of Good Practice

The third section is focused on implementation and reflection of initiatives for attracting young people for SET. These papers are present initiatives as well as evaluations even though their success and impact is often not directly measurable. They critically discuss under this topic the starting points and effects of so called inclusion measures aimed at changing the image of SET. Questions for discussion are: How are measures for motivating and attracting more young people for these subjects working? Which approaches do they use and what is the concept behind them? And finally, what methodological questions arise for evaluating such inclusion initiatives?

Frank-Stefan Becker in his paper, "Are young people lazy, blind or misguided? A fresh look at unpleasant facts", investigates the several reasons why small numbers of young people choose SET and argues that it could be rational in some way. Based on his analysis of the results of the NaBaTech-survey, his paper discusses the importance of image and status, the influence of society and peer groups, as well as financial rewards and career aspects of the engineering profession. He argues that, to a large extent, the universally observable trend away from SET (especially among young women) is due to understandable decisions. Society fails to provide sufficient visible "role models" of people who have succeeded as engineers rather than by switching from engineering to another profession. This is the case, he argues, not only for young women but as a general rule. As with other authors, Frank Becker thinks that the influence of the perceptions of young people, which translate into recognition and status among same-age peer groups, cannot be overestimated.

In "Mind the gap: Science and young people", Federica Manzoli and her colleagues Flora Di Martino, Danile Gouthier and Donato Ramani discuss the European project Gapp (Gender Awareness Participation Process, http://www.gendergapp.eu/). This project came out of statistics regarding young people and the study of science in Europe, where it was found that young people are losing contact with science and the number of science students in universities has decreased. In the research project, a sample of high school students and professionals engaged in academic or other science-related careers were interviewed, with a particular focus on girls. Central research questions were: How and where do the gender differences come in to being in science and technologies careers? What ideas influenced young people's choice regarding their professional future and how do the perceptions of science careers affect interest, motivation and subject choice at school, at the university, and consequently in their career? In which ways do girls and boys differ regarding these questions? Fields explored were the subject areas of mathematics, physics, IT, engineering, and chemistry. The methodological approach used involved 48 focus group discussions with students, teachers and parents in six European countries (Italy, Belgium, the Netherlands, Denmark, Portugal, Poland) and 60 in-depth interviews with opinion leaders in the field of S&T (scientists at the top of their careers, research institutes directors, consultants, experts in gender issues, involving women and men in equal proportion). The power of mass media is recognized as crucial in spreading a new way to see SET professions as appealing for youngsters and for girls in particular, while school and family remain the basic way to inform young people on the choice of a SET career. The opinions of experts show that the ability to become a good researcher is considered equal in men and women, while structural difficulties emerge for women compared with men at the higher levels of their career; SET professions require a strong commitment, which remains an obstacle in terms of the management of family life for women.

Barbara Roth talks in her paper, "Munich's gender-sensitive education – extend girls' world to the fascinating field of engineering", about examples of practical measures to allow girls to enter a profession in STEM (Science, Technology, Engineering, Mathematics). In her summary she sees mono-educative "girls-only" lessons or schools have the highest results in terms of motivating girls for STEM study and careers. Teachers and educators, starting from pre-school and upwards, should have a high-level of gender-knowledge in addition to practicing self-reflection and having awareness of their own gendered role. Technical education has to start very early, before socialisation influences kill girls' natural curiosity about nature and science. Gender-sensitive teaching aimed at motivating girls for STEM can counteract socialisation and should be inquiring, experiment orientated, including problem solving and practical examples from the daily life of the girls. Roth argues that on the other hand mandatory and reliable quotas at least 30% should be installed for male trainees in studies and vocational training: Child care assistants, male Kindergarten educators, and male primary school teachers. Diversity initiatives should be linked to accountability systems and tools to measure progress.

Sylvia Neuhäuser-Metternich and Sybille Krummacher in their paper on "Mentoring and video clips – ways against the threat of stereotyping" demonstrate examples of good practice in motivating young women for SET. Based on the phenomenon of how stereotypes threaten to negatively influence women's performance, the authors demonstrate two different but related approaches to changing the image of science and scientists and thus reducing the threat of stereotypes, mentoring and video clips and information material. Through the interaction with their mentees, educators and superiors are made aware of the multiple discouraging forces experienced by this group, including peer group attitudes, teaching methods and negative stereotypes on female performance in SET, and are led to revise their image of both the field and its actors (Sylvia Neuhäuser-Metternich/Sybille Krummacher 2009). Girls and young women already active in SET should therefore be the best advisors in developing effective strategies to recruit and retain female scientists. In the framework of the EU funded project UPDATE ("Understanding and Providing a Developmental Approach to Technology Education") engineering students at Dortmund University of Applied Sciences and Arts followed a multifunctional approach in taking part in the construction of video clips.

Marcel Lämmerhirt and Carmen Leicht-Scholten's paper on "Inspiring girls and boys in science, engineering and technology on virtual networks – the e-mentoring programme TANDEMkids at RWTH Aachen[14]" is based on schools' deficits in science subjects. The absence of technology lessons at schools, didactic deficits in the technical-scientific education, as well as the perpetual construction of gender-specific stereotypes concerning technology acceptance and competent technology application prevent equal access to technical-scientific subjects and professional education for girls and women. Their programme promises an individual, target group-specific and gender-specific access to MINT subjects (mathematics, informatics, natural sciences, technology) for pupils especially female pupils of 7th and 8th grade (age 12 to

---

14 The programme is funded by the Excellence Initiative at RWTH Aachen University.

13). Designed as e-mentoring, experienced mentors (students and Ph.D. candidates from the MINT subjects at RWTH Aachen University) support less experienced mentees (nationwide male and female pupils) for one year. The Mentoring programme is consecutively evaluated concerning the attitudes towards MINT subjects, expectations concerning the programme, and interaction and effectiveness in the mentoring relationship is also carried out online. The research methodology involved questionnaires, guide-supported interviews and log file analyses of the web-based communication forms.

Another university project, entitled "Mentoring Programme TANDEMschool – encouraging young people to study SET subjects by a coherent MINT cooperation concept at RWTH Aachen University", is presented by Gehrt Hartjen and Carmen Leicht-Scholten. The authors believe that mentoring as well as motivating pupils to study a scientific or technical (SET) subject (or at least get more attracted to the area) can also provide the participating students with important soft skills. The mentoring programme TANDEMschool addresses pupils of grade 11 to 13 (age 16 to 18) and offers them an individually chosen personal mentor (students and graduate students from the SET area at RWTH Aachen University). The MINT programme comprises of two mentoring programmes (TANDEMschool and TANDEMkids), a summer and winter school for pupils, university and doctorate scholarships on didactics as well as gender and diversity aspects in SET.

6 *MOTIVATION results*

The papers on MOTIVATION are summed up by an article from Felizitas Sagebiel et al.[15] about "How to change stereotypical images of science, engineering and technology (SET)? Results and conclusions from the European Project MOTIVATION". Teams from Austria, France, Germany, Slovak Republic, Spain, Sweden and The Netherlands studied the research questions: What images of disciplines and professions in science, engineering and technology (SET) are prevalent among adolescents, especially girls, and what factors influence them? The new focus of MOTIVATION was on general images of SET in youth magazines and television, in school and by the young people themselves taking initiatives of good practice on a national basis into account (Sagebiel 2008). Methods included document and media analysis, interviews, focus groups and drawings.

---

15 The original article has been published as Sagebiel, Felizitas et al. 2009. How to change stereotypical images of science, engineering and technology (SET)? Results and conclusions from the European Project MOTIVATION. Soziale Technik 4: 17-19. It is reprinted here with permission of Peter Wilding from Soziale Technik as responsible editor.

References

Abele, A.E. 2002. Geschlechterdifferenz in der beruflichen Karriereentwicklung. Warum sind Frauen weniger erfolgreich als Männer, in *Frauen machen Karriere in Wissenschaft, Wirtschaft und Politik: Chancen nutzen – Barrieren überwinden*, edited by B. Keller & A. Mischau. Nomos Verlag, Baden-Baden: 49-63

Baumert, Jürgen; Lehmann, Wilfried, et al. 1997. TIMSS – Mathematisch-Naturwissenschaftlicher Unterricht im internationalen Vergleich. Opladen

Becker, Frank-Stefan. 2010. Why don't young people want to become engineers? Rational reasons for disappointing decisions, in *European Journal of Engineering Education*, 35(4): 349-366

Beinke, Lothar. 2004. Berufsorientierung und peer-groups und die berufswahlspezifischen Formen der Lehrerrolle. Bad Honnef: Verlag K.H. Bock

Beinke, Lothar. 2006. Der Einfluss von Peer Groups auf das Berufswahlverhalten von Jugendlichen, in Übergang Schule und Beruf. Aus der Praxis für die Praxis – Region Emscher-Lippe. Wissenswertes für Lehrkräfte und Eltern, edited by Nikolaus Bley & Marit Rullmann. Recklinghausen: 249-265

Bos, Wilfried; Bonsen, Martin; Baumert, Jürgen; Prenzel, Manfred; Selter, Christoph & Walther, Gerd. 2008. TIMSS 2007. Mathematische und naturwissenschaftliche Kompetenz von Grundschülern in Deutschland im internationalen Vergleich. Münster: Waxmann

Buchmayer, Maria. 2008. Geschlecht lernen. Gendersensible Didaktik und Pädagogik. Innsbruck: StudienVerlag

Corneliussen, Hilde. 2002. The power of discourse – the freedom of individuals: Gendered positions in the discourse of computing. Department of Humanistic Informatics. Bergen, University of Bergen. positions in the discourse of computing. Department of Humanistic Informatics. Bergen, University of Bergen

Dahmen, Jennifer. 2006. Ergebnisse eines EU-Forschungsprojekts zur Situation von Studentinnen in den Ingenieurwissenschaften, Journal Netzwerk Frauenforschung NRW, 20: 36-42

Davies, S.; Kearnes, M. & Macnaghten, P. 2008. Nanotechnology and public engagement: a new kind of (social) science? In: http://dro.dur.ac.uk/6397/1/6397.pdf

Deaux, Kay & LaFrance, Marianna. 1998. Gender, in *The Handbook of Social Psychology*, edited by Daniel Todd Gilbert, Susan T Fiske & Lindzey Gardner. Boston et al: McGraw-Hill: 788-827

Dimbath, Oliver. 2007. Die (Be-)Deutung schulischer Berufsorientierung. Eine Analyse des Einflusses von Lehrerinnen und Lehrern auf die Berufswahl, in Bildung und Berufsorientierung. Der Einfluss von Schule und informellen Kontexten auf die berufliche Identitätsentwicklung, edited by Heike Kahlert & Jürgen Mansel. Weinheim: Juventa, 163-184

Dostal, Werner & Troll, Lothar. 2002. Die Berufswelt im Fernsehen. Folgen für das Berufsverständnis und den Berufswahlprozess, in *IAB-Studie 2002*, http://doku.iab.de/ibv/2004/ibv2404_57.pdf

Duru-Bellat, Marie. 2004. L'école des filles: Quelle formation pour quels rôles sociaux? L'Harmattan

Eccles, Jacquelynne S. 1994. Understanding women's educational and occupational choices: Applying the Eccles et al. model of achievement related choices. Psychology of Women Quarterly, 18: 585-609

European Commission. 2012. Meta-analysis of gender and science research. Synthesis report. Ed by Caprile, Maria et al. European Union

Faulkner, Wendy. 2000. 'Dualisms, hierarchies and gender in engineering', Social Studies of Science, 30(5): 759-92

Faulstich-Wieland, Hannelore. 2004. Mädchen und Naturwissenschaften in der Schule. Expertise für das Landesinstitut für Lehrerbildung und Schulentwicklung Hamburg. In: http://www.erzwiss.uni-hamburg.de/Personal/faulstich-wieland/Expertise.pdf

Fine, Cordelia. 2010. Delusions of Gender: How our minds, society and neurosexism create difference. New York: WW Norton

Götz, Maya. 2003. Was suchen und finden Mädchen in Daily Soaps? In: *Medien Sozialisation Geschlecht. Fallstudien aus der sozialwissenschaftlichen Sozialforschung*, edited by Renate Luca. München: KoPäd: 99-110

Gräßle, Kathrin. 2009. Wege ebnen für Frauen in technische Studiengänge. Leverkusen-Opladen: Verlag Barbara Budrich

Hanssen, Lucien. 2008. Ten lessons for a nanodialogue. The Hague: Rathenau Institute

Hannover, Bettina & Kessels, Ursula. 2004. Self-to-prototype matching as a strategy for making academic choices. Why German high school students do not like math and science. Learning and Instruction, 14(1): 51-67

Henwood, Flis, Gwyneth Hughes, et al. 2001. Cyborg lives in context: Writing women's technobiographies, in: *Cyborg Lives? Women's Technobiographies*, edited by F. Henwood, H. Kennedy and N. Miller. York, Raw Nerve Books: 11-34

Jim-Studie. 2006. Jugend, Information, (Multi-)Media. Basisuntersuchung zum Medienumgang 12- bis 19-jähriger. In: http://www.mpfs.de/fileadmin/JIM-pdf06/JIM-Studie_2006.pdf

Keller, Carmen. 1998. Geschlechterdifferenzen in der Mathematik. Zürich

Kessels, Ursula. 2005. Fitting into the stereotype: How gender-stereotyped perceptions of prototypic peers relate to liking for school subjects. European Journal of Psychology of Education, 20(3): 309-323

Kessels, Ursula & Hannover, Bettina. 2008. When being a girl matters less. Accessibility of gender-related self-knowledge in single-sex and coeducational classes. British Journal of Educational Psychology, 78(2): 273–289

Meulders, D.; Plasman, R.; Rigo, A. & O'Dorchai, S. 2010, Topic report. Horizontal and vertical segregation. Downoladed on 16/11/2010, available at: http://www.genderandscience.org/doc/TR1_Segregation.pdf

Mischau, Anina; Langfeldt, Bettina & Mehlmann, Sabine. 2009. Genderkompetenz als innovatives Element der Professionalisierung der LehrerInnenausbildung für das Fach Mathematik. Journal des Netzwerks Frauenforschung NRW, 25: 50-53

nanoBio-RAISE. Public Perceptions and Communication about Nanobiotechnology, available at: http://files.nanobio-raise.org/Downloads/NanoPublicFINAL.pdf

Neuhäuser-Metternich, Sylvia & Krummacher, Sybille. 2009. Ada Lovelace Mentoring – Engaging Girls and Women with Science and Technology, in *Science Education Unlimited. Approaches to Equal Opportunities in Learning Science*, edited

by Tanja Tajmel, & Klaus Starl. Münster, New York, München, Berlin: Waxmann: 169-178
Nissen, Ursula; Keddi, Barbara & Pfeil, Patricia. 2003. Berufsfindungsprozesse von Mädchen und jungen Frauen. Opladen: VS Verlag
Rasmussen, Bente. 1997. Girls and Computer Science: 'It's not me. I'm not Interested in Sitting Behind a Machine all day.' Women, Work and Computerization; Spinning a Web from Past to Future. Bonn: Springer Verlag
Rommes, Els; Overbeek, Geertjan; Scholte, Ron; Engels, Rutger & De Kemp, Raymond. 2007. 'I'm not interested in computers', Gender-based occupational choices of teenagers. Information, Communication & Society, 10(3): 299-319
Sagebiel, Felizitas. 2005. Attracting Women for Engineering. Interdisciplinary of Engineering Degree Courses in Mono-Educational versus Co-Educational Settings in Germany in Gender Equality in Higher Education, edited by Valeria Maione. Milano: Miscellanea Third European Conference Genoa, 13th-16th April 2003: 294-318
Sagebiel, Felizitas. 2007. Gendered organisational engineering cultures in Europe in Gender and Engineering – Problems and Possibilities, edited by Ingelore Welpe, Barabra Reschka & June Larkin. Frankfurt am Main: Peter Lang Verlag: 149-173
Sagebiel, Felizitas et al. 2008. Motivation of young people for studying SET. The gender perspective. In Proceedings of Sefi conference, Aalborg, 30th June -3rd July 2008, (CD). (Co-authors: Jennifer Dahmen, Bulle Davidsson, Anne-Sophie Godfroy-Genin, Els Rommes, Anita Thaler, Natasa Urbancíková)
Sagebiel, Felizitas et al. 2009. How to change stereotypical images of science, engineering and technology (SET)? Results and conclusions from the European Project MOTIVATION. Soziale Technik 4: 17-19
Sagebiel, Felizitas & Dahmen, Jennifer. 2005. Männlichkeiten in der europäischen Ingenieurkultur. Barrieren oder Aufforderung zur Anpassung für Frauen. Soziale Technik, 15(1): 19-21
Sagebiel, Felizitas & Dahmen, Jennifer. 2006. Masculinities in organisational cultures, in: *engineering education in Europe. Results of European project WomEng*. European Journal of Engineering Education, 31(1): 5-14
Sagebiel, Felizitas & Dahmen, Jennifer. 2010a. Attracting young people for studying SET under gender perspectives in Good practice to enhance the popularity and relevance of science education for scientific literacy. International PARSEL Conference, Berlin, 1st-4th March 2009, edited by Claus Bolte, Jack Holbrook & Wolfgang Gräber. Münster: Waxmann Verlag
Sagebiel, Felizitas & Dahmen, Jennifer. 2010b. Motivation – Die Gender Perspektive. Wie verändern wir das Image von Naturwissenschaft und Technik bei jungen Leuten? ADA-MENTORING, Fachzeitschrift für Mentoring und Gender Mainstreaming in Technik und Naturwissenschaften, 30: 4-15
Sagebiel, F. & Vázquez, S. 2010. Topic report. Stereotypes and identity. Downloaded on 16/11/2010, available at: http://www.genderandscience.org/doc/TR3_Stereotypes.pdf
Schreiner, Camilla & Sjöberg, Svein. 2006. How do students perceive science and technology? Science in School, Issue 1: 66-69
Steinke, Jocelyn. 2003. Media Images of Women Engineers and Scientists and Adolecent Girls. Conceptions of Future Roles. WEPAN Conference Proceedings, 8th-11th June 2003

http://dpubs.libraries.psu.edu/DPubS?service=Repository&version=1.0&verb=Disseminate&handle=psu.wepan/1181071742&view=body&content-type=pdf_1# (30.03.2009)

Thaler, Anita & Wächter, Christine. 2005. Conference Proceedings of the International Conference „Creating Cultures of Success for Women Engineers", 6th-8th October 2005. Leibnitz/Graz. Graz: IFZ Eigenverlag

Wächter, Christine. 2005. Interdisciplinary Engineering Education – An Opportunity for more Gender-Inclusiveness? In *Proceedings of the Fourth European Conference on Gender Equality in Higher Education*, 31st August-3rd September 2005, Oxford Brookes University. Oxford, (CD)

Wajcman, Judy. 1991, 1996. Feminism confronts technology. Cambridge: Polity Press

Wentzel, Wenka. 2005. Fünf Jahre Girls' Day – Mädchen Zukunftstag. Eine Zwischenbilanz. Zeitschrift für Frauenforschung und Geschlechterstudien, 25(1+2): 114-130

Zaidman, Claude. 1996. La mixité à l'école primaire. Paris: L'Harmattan, Coll. Bibliothèque du féminisme: 236

Ziegler, Albert; Kuhn, Cornelia & Heller, Kurt. 1998. Implizite Theorien von gymnasialen Mathematik und Physiklehrkräften zu geschlechtsspezifischer Begabung und Motivation. Psychologische Beiträge: 40 (3-4): 271-287

Zimmer, Karin et al. 2004. Kompetenzen von Jungen und Mädchen in PISA 2003. Der Bildungsstand der Jugendlichen in Deutschland – Ergebnisse des zweiten internationalen Vergleichs, edited by PISA-Konsortium Deutschland. Münster: 211-223

Zimmer, Rene; Domasch, Silke; Scholl, Gerd; Zschiesche, Michael; Petschow, Ulrich; Hertel, Rolf F. 2007. Nanotechnologien im öffentlichen Diskurs. Deutsche Verbraucherkonferenz mit Votum. Technikfolgenabschätzung – Theorie und Praxis, 16 (3): 98-101

Zoellner, Katharina. 2007. Jugendforen Nanomedizin. Chancen und Risiken, ethische und soziale Fragen der Nanomedizin aus der Sicht junger Erwachsener. In: http://www.nano-jugend-dialog.de/daten/Nanojugenddialog_endbericht_website.pdf

Socialization agents and the gendered choice of educational paths: Perpetuation or fragility of gender stereotypes?

Susana Vázquez-Cupeiro

This research has been funded by the European Commission, through the study 'Meta-analysis of gender and science research' (2008-2010), under the 7th RTD European Framework Programme commissioned by the DG Research to the consortium led by CIREM (Spain) and made up of: Université Libre de Bruxelles (Belgium), Inova Consultancy Ltd. (UK), Fondazione Giacomo Brodolini (Italy), Bergische Universität Wuppertal (Germany) and Politikatörténeti Intézet KHT (Hungary). This paper uses data from the thematic report 'Stereotypes and Identity', written by Felizitas Sagebiel and Susana Vázquez-Cupeiro.

Abstract

A broad array of explanations to understand the under-representation of women in science, technology, engineering and mathematics (STEM) has been put forward in the literature of recent decades. This review looks at the choice of educational and professional trajectories of young girls and boys through research focused on the socialization process. Giving that stereotypes play a significant role in the biased construction of gender roles, it is suggested that socialization agents are powerful not only in transmitting and perpetuating gender stereotypes but also in challenging them.

1 Introduction

Even if there are no remarkable differences between girls and boys in terms of their skills and attitudes (Hyde 2006; NAS 2007; Xie & Shuman 2003), their choice of subjects and careers continues to be gendered. It has been suggested that the reason for this gendered choice of educational paths is related to the fact that girls do not like science-related subjects or they do not see it as an integral part of their lives (Jones et al. 2000; Sadker & Sadker 1994; Schwartz & Hanson 1992). Further, Xie and Shauman (2003) have argued that most girls do not choose science-related degree courses and trajectories (while boys are not likely to choose 'women's fields'), because they do not perceive the field as being within their realm of opportunity. Why? Given the lack of results supporting the explanations that rely on sex-related biological abilities, a review of the main theoretical frameworks concludes that the gendered choice of different academic paths and professional careers is largely the result of socialization factors (Campbell, Verna & O'Connor-Petruso 2004; Eccles et al. 2000; Simpkins, Davis-Kean & Eccles 2005).

The socialization process involves the internalization of the social world in the context of a particular social structure. It is during primary socialization that children internalize the world filtered through the eyes of close adults and peers. Yet socialization accompanies us throughout life with the intervention of other key agents of socialization. Family (mainly parents), school (teachers), peers and mass media, and increasingly ICT (Information and Communication Technologies), play a key part in the transmission of gender roles and therefore help to define the 'appropriate' patterns of behavior. Yet, given the fact that the socialization agents are decisive in the choice of the academic and vocational itineraries of young girls and boys, and that these decisions are shaped by the transmission and perpetuation of gender stereotypes (Leaper and Friedman 2007), the socialization agents are also essential in terms of refuting them and promoting a more egalitarian perception of opportunity contexts for men and women.

2   The family's influence on boys' and girls' educational choices and paths

Family is the primary socialization agent. Parental perceptions and attitudes influence children's development and interests (for instance what to choose as a major). From early childhood education, and in line with gender stereotypes, parents tend to encourage sex differences in behavior and experience. This occurs not only by treating boys and girls differently but also by estimating their abilities differently (Jacobs & Eccles 1992; NAS 2007). Girls are expected to be better at language skills and boys at maths (Gutbezahl 1995; Halpern 2006). Other studies go further and suggest that many adults believe that boys have an innate mathematical ability and, as a result, parents (and teachers) tend to underestimate the mathematical intelligence levels of girls (Sadker & Sadker 1994). Moreover, the differentiated gender success is often explained on the basis of "innate" abilities in the case of sons and "effort to compensate lack of skills" in the case of daughters (Duru-Bellat 2005).

A body of literature has also found differences in the estimation of abilities on the part of fathers and mothers. Some studies suggest that fathers tend to have a less traditional conceptualization of gender roles than mothers, and that they are more likely to encourage girls to get involved in activities that are not traditionally associated with women (McHale et al. 2004). Eccles and Jacobs (1986) have demonstrated that the relationship is particularly strong with respect to mothers' perception of the difficulty of science for their child and the self-efficacy and interest of the child in the subject. Furthermore, mothers who endorse a male-maths stereotype tend to underestimate their daughters' ability in maths (Halpern 2006; Tenenbaum & Leaper 2003).

Parents may also inadvertently influence their child's lack of interest in science by responding differently to sons and daughters. Some research has found that parents are more likely to explain scientific concepts to sons than to daughters (Crowley et al. 2001). Also, parents use sex-differential language (e.g. parents tend to use less cognitively-demanding language with daughters) (Tenenbaum & Leaper 2003). Further, parents often buy science materials (computers, books, games, etc.) for boys more often than for girls (Simpkins, Davis-Kean & Eccles 2005). As a result, girls may be less likely to be encouraged to pursue science related studies (Xie & Shauman 2003), while more likely to underestimate their own mathematical abilities.

Some researchers agree that the socioeconomic status of the family of origin is a strong educational predictor (Education Sector 2006). Yet Xie and Shauman (2003) point out that, in spite of the strong association between the family's socioeconomic status (parental income) and maths/science achievement, this variable is unlikely to explain gender differences independently of the sex of the children. The authors suggest that parental attitudes and expectations of their children may be more likely related to their own level of education. Highly educated parents are more likely to expect their children to go to college and to be able to afford the expenses involved. Further, they may contribute to enhancing their children's scientific skills, an aspect that may have an effect on the participation, achievement and persistence of women in STEM (Science, Technology, Engineering and Mathematics). Other studies conclude that parents' interest and engagement in science and mathematics predicts the grades that children earn later on in school (Jacobs & Eccles 1992). In this sense, children's self-evaluation of academic competence appears to be more strongly related to their parents' appraisals of their academic ability than to their academic performance.

Social forces also affect gender differences in career choice through role modeling (Bandura 1986), as young males and females tend to make educational and occupational choices emulating same-sex adult experiences. Parents act as role models for their sons and daughters. So, the presence and/or absence of female role models (to identify with) in the family in STEM fields traditionally portrayed as masculine is crucial in understanding boys' and girls' divergent roads (AAUW 2000 & 2010; Håpnes & Rasmussen 2000; Xie & Shauman 2003). Suter (2006) suggests that the encouragement of the family is an important social resource particularly for girls, as female students in engineering and other branches of science often have at least one parent with a professional background in one of these disciplines. This finding highlights the importance of having a supportive network as well as a female/male role model within the family.

3 The role of school in gender-biased education

Throughout the process of secondary socialization, when more general values are attained, boys and girls choose subjects to study. The decision, in accordance with desirable gender characteristics, is taken under the increasing influence of new socialization agents: The school system, peer groups and the mass media. The school exerts influence through diverse channels: "the availability of courses, sporting facilities, and extracurricular activities; the quality of teaching, teacher's expectations, and guidance; the availability and orientation of guidance counselors and the characteristics of the student's peers" (Xie & Shauman 2003: 8). But the school is an institution with substantial masculine orientation and, as such, tends to reproduce the dominant culture and the existing social status quo (Bauer 1999; Schwartz & Hanson 1992). Classmates, teachers and their stereotypes strongly influence children's conceptions of what they can achieve (NAS 2007; Steele 1997). In fact, according to Wajcman (1991), "in modern societies it is the education system, in conjunction with other social institutions, which helps to perpetuate gender inequalities from generation to generation" (p.151).

In the literature that seeks to explain why girls choose science-related studies to a lesser extent than boys, there is a line of research that examines the pedagogies, teaching styles and classroom interactions between teachers and students. It has been suggested that a crucial reason explaining the gaps between sexes is the (often unaware) teachers' stereotyped (sexist) attitudes towards students in the classrooms (Aksu 2005). Male and female teachers live in a social context and therefore assume that some disciplines are more feminine than others (Vendramin et al. 2003) or that boys are better at maths (Sadker & Sadker 1994). So, it has been suggested that teachers participate in the consolidation of existing gender stereotypes and the activation of "self-fulfilling prophecies" (Sáinz & González 2008).

A recurring theme regarding the cause of the possible gender divide in science related disciplines comes from the type of classroom instruction boys and girls receive in primary school, high school and college. In spite of the fact that gender bias in the interaction between the teacher and students was found in all subject areas, the greatest bias was found in the math and science classrooms. On the basis of their stereotyped expectations, whether consciously or unconsciously, teachers often treat girls and boys differently (Gutbezahl 1995). For instance, some research provides empirical evidence that boys received more attention, challenging interaction and constructive feedback from teachers than girls did (Smith 1996). Other authors have gone further and suggest that while boys are encouraged to participate, are given more eye contact and longer wait times, girls are the "invisible members of classrooms" (Bauer 1999; Schwartz & Hanson 1992). Other research has also identified inequalities in classroom instruction. Boys are called on more often, interact more with the teacher, are asked complex, higher order and open-ended questions more often, are called on to use abstract reasoning more often, and tend to dominate the classroom. The conclusion: Unbalanced gendered instruction works to the detriment of the maths and science performance of female students (Sadker 1999).

According to Sáinz and González (2008), the negative influence of teachers and academic advisors, the male-biased technology curriculum, together with the small number of women teachers who serve as role models to younger students form the main arguments used to explain the segregation of women in SET. Moreover, as stated by Deaux and Major (1987), gender stereotypes are also made more salient in environments where the presence of women is the exception. As a result, many girls end up choosing specialities closer to the female gender role within science-related fields.

4   Friends, peer groups and leisure

Friends and peer groups are the main reference point for teenagers. The peer-culture and the opinion of friends influence the choice of studies, leading to the selection of those studies understood as "normal", according to the assigned social roles (Vendramin et al. 2003), and that facilitate continuity in the group. Some research has found that students, particularly girls, feel that social pressures have an impact on their achievement, and in this sense an even stronger influence than parents and family are their peers (Bryan 1997). Related to the fact that girls seem to be more sensitive than boys to the perception of social acceptance and to the need to be accepted in the group (Håpnes & Rasmussen 2000), is the notion that "females camouflage talent" (Campbell, Verna & O'Connor-Petruso 2004). This is based on the hypothesis that while during preschool and primary school years gifted females are encouraged to develop their talents, during early adolescence and adulthood many gifted females learn to camouflage them in an effort to gain acceptance by other females and by males, for dating and marriage. As a result, their career development is limited (Kerr 1994).

During their leisure time children and teenagers are also influenced by gender-related stereotypes that may predispose them to make certain career choices and not others. In general, youth culture is not favorable to technical subjects (seen as more difficult) and this perception may be decisive when it comes to choosing the field of study, first at school and later at university. Thus, gender gaps in performance might also be the result of different "outside-of-school" experiences of males and females (peers, media, games, the new technologies, etc.). Such differences in experiences and activities might in turn lead to differences in the motivation to seek knowledge about science, which in turn will lead to performance differences in different content areas of science. In summary, the gender system and social gender roles channel women into more feminine careers and professions and distance them from SET and ICT educational and career patterns (Margolis & Fischer 2003). As a result, it seems difficult to separate "preference" from historical patterns of access, which are strongly gendered.

5 Mass Media and the polarization of gender roles

The mass media and ICT are nowadays two of the most powerful channels of youth socialization. The importance of these instruments lies not only in the content they transmit, but also in the new ways of establishing relationships, transmitting information and communicating. According to Bandura (1969), children learn gender stereotypes from media sources that, in turn, influence their attitudes and behaviors. This happens through repeated observations ("identificatory learning") of both actual models in their social environments (parents and teachers) and symbolic models (images and characters they encounter in the media). The mass media, in particular television, perpetuates traditional gender stereotypes in that they reflect dominant social values. The portrayal of both men and women (adverts, programs, soap operas, films, etc.) is largely traditional and stereotyped and this serves to promote a polarization of gender roles. In fact, as suggested by Steinke (2005), before children reach adolescence, when most begin to develop individual identities and prepare for future roles, they are likely to have seen countless media images that emphasize gender qualities and urge conformity to traditional stereotypes.

It has been found that girls and young women develop gender schemas that lead them to label certain high-status professions (technology, scientists, engineering) as masculine and, as a result, to restrict their professional aspirations (Steinke 2005). Images of male and females interactions reinforce traditional social and cultural assumptions about the role of women in science through overt and subtle forms of stereotyping. Moreover, some researchers argue that during the middle and high school years, interest in science declines (AAUW 2000). Thus, it is during adolescence, when girls and boys show awareness of gender roles and the media become influential sources of information that they first seem to lose interest in these subject areas. For this reason, Steinke (2005) suggests that there is a connection between cultural representations of gender and the gender gap.

Research also indicates that the media (in particular television) is a major source of information about scientists for middle school-aged children. The literature exploring the images associated with science and scientists has documented stereotypic portrayals. As a result, students perceive scientists as predominantly (white) males with glasses, lab coats and facial hair (Barman 1997). Further, a study examining gender stereotyping in mass-media portrayals of male and female scientists on television programs likely to be viewed by middle school-aged children (dramas, cartoons, situation comedies and scientific educational programs) concludes: 1) male scientist characters are more prevalent than female scientist characters; and 2) while male scientists show masculine attributes (independence and dominance), female scientists are portrayed with feminine attributes (dependence, caring, and a romantic nature) (Steinke et al. 2008). Furthermore, research on middle school-aged children found that boys who indicated the media was very important had more negative attitudes towards women in science. The findings are important not only because this may affect girls' behavior, but also because it may affect their adult views.

6   Gender stereotypes: An unavoidable reality?

Throughout childhood and into adolescence, children develop specific views about gender roles according to the society in which they live. Gender roles are socially constructed and vary according to the culture and over the course of people's lives. Children not only learn about gender roles from their parents, teachers and peers, but also from media sources. Given that most of the behavior associated with gender is learned rather than innate and that stereotypes are not fixed, the question is: How can gender stereotypes be challenged?

Negative gender stereotyping of abilities can strongly influence children's conceptions of what they can achieve and their future career choices (Eccles et al. 2000; Steele 1997). According to the meta-analysis developed by Leaper, Anderson and Sanders (1998), the extent of the influence of parents' gender beliefs began to weaken after the mid-1980s, as a result of the diminishing weight of gender stereotyping. In this sense, it has been demonstrated that the "nature" of the family (traditional versus egalitarian families) may be very influential in determining the nature of the children's socialization process as well as the choice of studies. Some research shows that children living in households characterized by gender equality tend to make less stereotyped classifications of occupations, while girls from such families want to pursue non-traditional careers more often

(Weisner & Wilson-Mitchell 1990). In relation to gender roles, girls from these households tend to express a stronger interest in mathematics (Jacobs & Eccles 1992; NAS 2007) and obtain better results in secondary school, especially in math and science (Updegraff, McHale & Crouter 1996). Further, in more egalitarian families, aware of the lower expectations for girls, parents try to adjust their attitudes, aspirations and behavior (Jacobs & Eccles 1992) accordingly. By contrast, more traditional environments are more likely to hold the belief that boys find science easier and more interesting than girls do and when it comes to SET disciplines, parents not only encourage but also expect better achievement and greater persistence from boys (Eccles & Jacobs 1986). This suggests that gender equality, mainly in the family, but also in the educational context, can mean an opportunity to reduce gender gaps and to encourage boys and girls to choose their professions freely. As a result, even if the results are not conclusive, parents should be "educated" to avoid the reproduction of gender stereotypes, promote higher self-esteem and self-confidence, and encourage girls to study, persevere and excel in math and science (Sadker 1999).

The "why" and the "what can be done about it" in the school system, in relation to the gendered choice of educational paths, have been the object of debate mainly in Anglo-Saxon, Nordic and Continental countries. According to the report 'Gender Differences in Career Choices: Why Girls Don't Like Science' (CCL 2007), if cultural and environmental factors, rather than biological predispositions, account for the gender gap in science, young girls' disengagement with science can be prevented and their natural interests fostered. Thus, several steps have been suggested to foster girls' interest in science as well as a number of programs to encourage girls to pursue studies and careers in SET.

Some of the suggestions made by researchers are related to teachers. For example, that they should improve the way they conduct the classrooms (e.g. reduce speed and give students more time) (Bauer 1999), provide balanced instruction (e.g. provide equal opportunities for boys and girls while respecting their differences), change their teaching methods (e.g. help girls to see how math fits into their lives as well as female students with special needs) (Roivas 2009), promote single-sex mathematics classes (Dunlap 2002), single-sex classes or interdisciplinary courses (Sagebiel 2005), send positive messages to build students' confidence (Bauer 1999), use cooperative learning groups rather than competitive ones (Schwartz & Hanson 1992), provide positive (female) role models in the areas of math and science and, among other things, be trained in the area of gender inequities in mathematics and science (Levi 2000). Most of the debate, however, has been centered on aspects related to roles models, single-sex classes, pedagogic/learning environments and instruction in gender equality for teachers. A body of literature defends the need to expose girls from an early age to female role models through pioneering programs to encourage the interest in SET careers (Dunlap 2002; Margolis & Fisher 2003). Certain studies have, nonetheless, examined and questioned the role model argument, especially in relation to its claims regarding the benefits of matching teachers and learners by gender. For instance, research carried out in Northern European countries concludes that high school students attach relatively little importance to the teacher's gender (Butler & Christiansen 2003). In fact, rather than gender, students place greater value on the teachers´ capacity to impose discipline in the classroom in a friendly, sensitive and impartial way. Yet, most studies recognize the need not only to make teaching a more inclusive profession but also to increase the availability of role models to break down enduring gender stereotypes (Mulholland & Hansen 2003).

There is also much debate amongst researchers today about the value and possible benefits of single-sex education. The advocates argue that single-sex education can improve motivation, behavior and achievement (SEED 2006). It allows girls and boys greater freedom to choose subjects not associated with their gender and benefits girls' self-esteem and participation while stimulating their interest in mathematics/science classes (Younger & Warrington 2007). By contrast, opponents of single-sex education suggest that, despite some success stories, no conclusive evidence has documented higher achievement for girls in single-sex classes and colleges than in coeducational institutions (AAUW 2002; Sanders & Peterson 1999; Skelton & Francis 2009; Smith 1996).

Some proposals emphasize the need to bring the gender perspective to bear in the study of the transmission of knowledge with respect to power relations and the different forms of inequality. It has been suggested that gender should be understood as a theoretical analytical tool (not linked to fixed characteristics attributed to girls and boys). In this sense, Schiebinger et al. (2010) point out that, while incorporating gender methods and analysis into the SET curriculum, there is also an imperative to train scientists, engineers and policymakers in order to implement this approach. Moreover, while some investigations stress the need to improve instruction for faculty members and councilors as to how to combat gender inequity in the classroom (Hyde 2006), other research focuses on inclusive pedagogies and teaching environments and styles that foster greater motivation for girls towards mathematics and technologies (Paechter 2003). Accordingly, it has been argued that "feminizing" the channels of transmission of knowledge (for instance, through approaches committed to creating more feminized teaching styles), should be eluded in order to avoid feeding into the dichotomous stereotypes according to which girls and boys are "suited" to different subjects (Mendick 2005).

The media and ICT are powerful agents for transmitting gender roles and pervasive stereotypes. While some studies have documented the negative effects of media images of scientists on children's attitudes towards science, others have noted the positive effects of these images. Although images of women in the media have improved in recent years, many researchers examining media content document stereotypical images that reinforce traditional conceptions of femininity and masculinity, a fact that, following Ruvolo and Markus (1992), may limit adolescents' visions of "possible selves". The media are inevitably socializing children into traditional stereotypical roles because of the prevalence of such images on TV and the importance ascribed to them by children (Sharpe 1976). But the media also offer a wide range of potential role models, both positive and negative. As a result, while there is little doubt that the media presents largely traditional gender images, it can also be used to promote a gender-bias-free socialization.

Over time, significant changes have taken place. The old dichotomy between arts and humanities (female-dominated) and science (male-dominated) no longer seems to hold (with the exception of engineering and ICT). Sweeping changes in the gender system have occurred. For example, it has been noted that gender disparities in terms of subject choice and career destination have decreased in some countries: 1) for (some) middle-class students (e.g. those attending fee-paying schools) (Arnot et al. 1999) and 2) among girls and boys attending single-sex schools (where there seems to be less pressure to conform to sex stereotypes) (Skelton and Francis 2009). Focusing on gifted girls, some researchers suggest that sociocultural changes occurring over the past three decades have gradually resulted in some changes in women's attitudes towards career choices (Leung, Conoley & Scheel 1994). In fact, the top career choices for gifted early-adolescent males and females seem to be identical (Reis, Callahan and Goldsmith 1996).

Furthermore, while engineering and ICT continue to be male-dominated areas (Castaño 2008; Castaño et al. 2011; Cockburn 1985; Faulkner 2001; Marcelle 2000; Vázquez 2010; Wajcman 1991), medicine is now gender-balanced in many countries and is even female-dominated in others (Kilminster et al. 2007), and nursing is becoming an increasingly masculinized field (Burton & Misener 2007; Chusmir 1990; Hayes 1989; Lemkau 1984; Neighbours 1999; Williams 1992). These changing trends, not always acknowledged in all recent literature, clearly reveal an emerging 'fragility' of gender stereotypes when it comes to certain disciplines and professions.

7 Some conclusions

Socialization agents play a significant role in the transmission of gender stereotypes and, as a result, in the gendered choice of the educational and professional paths of young girls and boys. The fact that women are a minority in SET has a number of serious implications. First, girls are denied the right to develop their full potential. This is not only important in terms of gender equity, the waste of their intellectual potential and talent also has negative effects in terms of socio-economic development. Second, an argument that is gaining more weight and which has recently been noted by the European Commission, is the unmet demand for professionals (particularly of women) in the system of science and technology. Last, but not least, from an epistemological point of view the development of new scientific perspectives may be only encouraged and enriched (quality and excellence) through diversity (gender, ethnic, etc.). These implications underline the need to challenge gender stereotypes in order to contest the gendered distribution of scientific areas.

Much effort is needed, at multiple levels, to ensure that males and females are free to develop their talents and to pursue the careers best suited to their own personal interests and desires. Besides a perspective that assumes that girls do not "fit" well and should "masculinize" themselves in order to adapt better, a considerable part of the literature emphasizes the need to transform the contexts of learning in order to make them more inclusive of girls (Caprile et al. 2012). In this sense, Schiebinger (2008 & 2011) suggests that besides implementing measures focusing on "women", it is necessary to concentrate on "institutions" while questioning the very nature of "knowledge" (for instance by promoting gender analysis). Only then we will be closer to challenging the current gender stereotypes which lead girls and boys to a gendered choice of educational and professional trajectories.

It has been shown how the family, school, peers and mass media/ICT are decisive agents in the construction of gender roles and, as a result, in the choice of the academic and vocational preferences of young girls and boys. Given that this process is shaped by gender stereotypes, the socialization agents are also powerful in challenging them, promoting a gender-bias-free socialization and, as a result, favoring the equal participation of women and men in SET. How can this be done? By becoming aware of the persistence of gender stereotypes; by being aware of its effects during the socialization process; by acknowledging the need to challenge them; and by compromising in the effort to become more egalitarian from the inside-out.

References

Aksu, B. 2005. Barbie Against Superman: Gender Stereotypes and Gender Equity in the Classroom. Journal of Language and Linguistic Studies, 1(1), April

American Association of University Women (AAUW). 2000. Tech-savvy: Educating girls in the new computer age. Washington, D.C. AAUW Education Foundation. In: http://www.aauw.org/2000/techsavvybd.html [10.1.2013]

American Association of University Women (AAUW) (2002) Single-sex education. In: www.aauw.org/1000/pospapers/ssedbd.html [10.1.2013]

American Association of University Women (AAUW) (2010) Why So Few? Women in Science, Technology, Engineering, and Mathematics. In: http://www.aauw.org/learn/research/whysofew.cfm [10.1.2013]

Arnot, M., David, M. & Weiner, G. 1999. Closing the Gender Gap: Postwar Education and Social Change. Cambridge: Polity Press

Bandura, A. 1969. Social-learning theory of identificatory processes, in Handbook of socialization theory and research, edited by D. A. Goslin. Chicago: Rand McNally: 213-262

Bandura, A. 1986. Social foundations of thought and action: A social cognitive theory. Englewood Cliffs NJ: Prentice-Hall

Barman, C. R. 1997. How Do Students Really View Science and Scientists?, Science and children, Sept: 30-33. In: http://www.eiu.edu/~scienced/329options/crbscience.html [10.1.2013]

Bauer, K. 1999. Promoting gender equity in schools, Contemporary Education, 71: 22-25

Bryan, C. 1997. Gender differences found in the way boys and girls solve math problems. In: http://www.apa.org/releases/math2.html [10.1.2013]

Burton, D. A. & Misener, T. R. 2007. Are you man enough to be a nurse? Challenging male nurse media portrayals and stereotypes, in *Men in nursing: history, challenges and opportunities*, edited by C.E. O'Lynn & R. E. Tranbarger. New York: Springer Publishing Company: 255-270

Butler, D. M. & Christensen, R. 2003. Mixing and Matching: The Effect on Student Performance of Teaching Assistants of the Same Gender, Political Science and Politics, 36: 781-786

Campbell, J. R.; Verna, M. & O'Connor-Petruso, S. 2004. Gender paradigms. Paper presented at the IRC-2004 Conference, Lefkosia, Cyprus. In: http://www.iea.nl/fileadmin/user_upload/IRC2004/Campbell_Verna_OConnor-Petruso.pdf [10.1.2013]

Canadian Council of Learning (CCL). 2007. Gender differences in career choices: Why girls don't like science. Lessons in Learning. November. In: http://www.ccl-cca.ca/pdfs/LessonsInLearning/Nov-01-07-Gender-Difs.pdf [10.1.2013]

Caprile, M. (coord.); Addis, E.; Castaño, C.; Klinge, I.; Larios, M.; Meulders, D.; Müller, J.; O'Dorchai, S.; Palasik, M.; Plasman, R.; Roivas, S.; Sagebiel, F.; Schiebinger, L.; Vallès, N. & Vázquez-Cupeiro, S. 2012. Meta-analysis of gender and science research. Synthesis report. Directorate-General for Research and Innovation. Brussels: European Commission. In: http://www.genderandscience.org/doc/synthesis_report.pdf [10.1.2013]

Castaño, C. (ed). 2008. La segunda brecha digital. Madrid: Ediciones Cátedra

Castaño, C.; González, A., Müller, J.; Sáinz, M.; Vázquez, S.; Vergés, N.; Palmen, R. & Rodríguez del Barrio, A. 2011. Quiero ser ingenier@. Barcelona: Ediciones UOC

Chusmir, L. H. 1990. Men who make nontraditional career choices, Journal of Counseling & Development, 69: 11-16

Cockburn, C. 1985. Machinery of dominance: Women, men and technical know-how. London: Pluto

Crowley, K.; Callanan, M. A.; Tenenbaum, H. R. & Allen, E. 2001. Parents explain more often to boys than to girls during shared scientific thinking, Psychological Science, 12: 258-261

Deaux, K. & Major, B. 1987. Putting gender into context: An interactive model of gender related behavior, Psychological Review, 94: 369-389

Dunlap, C. E. 2002. An Examination of Gender Differences in Today's Mathematics Classrooms: Exploring Single-gender Mathematics Classrooms. Ohio: Cedarville University

Duru-Bellat, M. 2005. L'école des filles: Quelle formation pour quels rôles sociaux? Paris: L'Harmattan

Eccles, J. S. & Jacobs, J. E. 1986. Social forces shape math attitudes and performance, Signs: Journal of Women in Culture and Society, 11(21): 367-380

Eccles, J. S.; Frome, P.; Suk Yoon, K.; Freedman-Doan, C. & Jacobs, J. 2000. Gender-role socialization in the family: a longitudinal approach, in *The Developmental Social Psychology of Gender*, edited by T.Y. Eckes & H. M. Trautner. New Jersey: Lawrence Erlbaum Associates Publishers: 333-360

Education Sector. 2006. The Truth About Boys and Girls. Washington, DC: Education Sector

Faulkner, W. 2001. The technology question in feminism: a view from feminist technology studies, Women's Studies International Forum, 24 (1)

Gutbezahl, J. 1995. How negative expectancies and attitudes undermine female's math confidence and performance: A review of the literature, ERIC/SCMEE database. In: http://eric.ed.gov/ERICDocs/data/ericdocs2sql/content_storage_01/0000019b/80/13/b8/97.pdf [10.1.2013]

Halpern, D. F. 2006. Biopsychosocial contributions to cognitive performance. Panel 1: Cognitive and Biological Contributions. Washington, DC: The National Academies Press. In: http://www.ncbi.nlm.nih.gov/bookshelf/picrender.fcgi?book=nap11766&blobtype=pdf [10.1.2013]

Håpnes, T. & Rasmussen, B. 2000. New technology increasing old inequality? In *Women, Work and Computerization: Charting a Course to the Future*, edited by E. Balka, E. & R. Smith. Canada: British Columbia

Hayes, R. 1989. Men in female-concentrated occupations, Journal of Organizational Behavior, 10: 201-212

Hyde, J. S. 2006. Gender Differences and Similarities in Abilities. Panel 1: Cognitive and Biological Contributions. Washington, DC: The National Academies Press. In: http://www.ncbi.nlm.nih.gov/bookshelf/picrender.fcgi?book=nap11766&blobtype=pdf [10.1.2013]

Jacobs, J. E. & Eccles, J. S. 1992. The influence of parent stereotypes on parent and child ability beliefs in three domains, Journal of Personality and Social Psychology, 63(6): 932-44

Kerr, B. 1994. Smart girls: A new psychology of girls, women and giftedness. Scottsdale: Gifted Psychology Press

Kilminster, S.; Downes, J.; Gough, B.; Murdoch-Eaton, D. & Roberts, T. 2007. Women in medicine – is there a problem? A literature review of the changing gender composition, structures and occupational cultures in medicine, Medicine Education, Jan, 41(1): 39-49

Leaper, C. & Friedman, C. K. 2007. The socialization of gender, in *Handbook of socialization. Theory and research*, edited by J.E. Grussec & P.D. Hastings. New York: Guilford Press: 561-587

Leaper, C., Anderson, K. J. & Sanders, P. 1998. Moderators of gender effects on parents' talk to their children: a meta-analysis, Developmental Psychology, 34(1): 3-27

Lemkau, J. P. 1984. Men in female-dominated professions: Distinguishing personality and background features, Journal of Vocational Behavior, 24: 110-122

Leung, S. A.; Conoley, C. W. & Scheel, M. 1994. Factors affecting the vocational aspirations of gifted students, Journal of Counseling and Development, 72: 298-303

Levi, L. 2000. Gender equity in mathematics education, Teaching Children Mathematics, October: 101-105

Loughrey, M. 2007. Just how male are male nurses?, Journal of Clinical Nursing, 17(10): 327-1334

Marcelle, G. M. 2000. Transforming Information & Communications Technologies for Gender Equality. Gender in Development Programme, UNDP

Margolis, J. & Fischer, A. 2003. Unlocking the clubhouse: Women in computing. Cambridge, MA: The MIT Press

McHale, S. M.; Shanahan, L.; Upedegraff, K. A.; Crotuer A. C. & Booth, A. 2004. Developmental and individual differences in girls' sex-typed activities in middle childhood and adolescence, Child development, 75(5): 1575-1593

Mendick, H. 2005. A beautiful myth? The gendering of being/doing 'good at maths', Gender and Education, 17(2): 203-219

Mulholland, J. & Hansen, P. 2003. Men who become primary teachers: an early portrait, Asia-Pacific Journal of Teacher Education, 31(3): 213-224

National Academy of Sciences (NAS). 2007. Beyond Bias and Barriers: Fulfilling the Potential of Women in Academic Science and Engineering, Washington, D.C.: Committee on Science, Engineering and Public Policy (COSEPUP). In: http://www.nap.edu/openbook.php?record_id=11741 [10.1.2013]

Neighbours, C. 1999. Male nurses, men in a female dominated profession: the perceived need for masculinity maintenance. In: http://studentnurse.tripod.com/men.html [10.1.2013]

Paechter, C. 2003. Power/knowledge, gender and curriculum change, Journal of Educational Change, 4: 129-148

Reis, S. M.; Callahan, C. M. & Goldsmith, D. 1996. Attitudes of adolescent gifted girls and boys towards education, achievement, and the future, in *Remarkable women: Perspectives on female talent development*, edited by K. D. Arnold, K. D. Noble and R. F. Subotnik. Cresskill: NJ. Hampton Press: 209-224

Roivas, S. 2010. Meta-analysis of gender and science research – Country group report Nordic countries. In: http://www.genderandscience.org/doc/CGR_Nordic.pdf [10.1.2013]

Ruvolo, A. P. & Markus, H. R. 1992. Possible selves and performance: The power of self-relevant imagery, Social Cognition, 10(1): 95-124

Sadker, D. 1999. Gender equity: still knocking on the classroom door, Educational Leadership, 4: 22-26

Sadker, M. & Sadker, D. 1994. Failing at fairness: how our schools cheat girls. New York: Simon & Schuster

Sagebiel, F. 2005. Attracting Women for Engineering: Interdisciplinary of Engineering Degree Courses in Mono-Educational versus Co-Educational Settings in Germany" in *Gender Equality in Higher Education*, edited by V. Maione, Milano: Miscellanea Third European Conference Genoa: 294-318

Sagebiel, F. & Vázquez-Cupeiro, S. 2010. Meta-analysis of gender and science research – Topic report Stereotypes and identity. In: http://www.genderandscience.org/doc/TR3_Stereotypes.pdf [10.1.2013]

Schiebinger, L. 2008. Knowledge Issues, in Gendered Innovations in Science and Engineering, edited by L. Schiebinger. Stanford: Stanford University Press: 1-21

Schiebinger, L.; Klinge, I.; Arlow, A. & Newman, S. 2010. Gendered Innovations. Mainstreaming sex and gender analysis into basic and applied research. Meta-analysis of gender and science research – Topic report. In: http://www.genderandscience.org/doc/TR6_Content.pdf [10.1.2013]

Schiebinger, L. & Schraudner, M. 2011. Interdisciplinary Approaches to Achieving Gendered Innovations in Science, Medicine, and Engineering, Interdisciplinary Science Reviews, 36(2): 154-67

Sáinz, M. & González, A. M. 2008. La primera y segunda brecha digital: Educación e investigación, in La segunda brecha digital, edited by C. Castaño. Madrid: Ed. Cátedra

Sanders, J. & Peterson, K. 1999. Close the gap for girls in math-related careers, The Educational Digest, 12: 47-49

Schwartz, W. & Hanson, K. 1992. Equal mathematics education for female students, ERIC/CUE Digest, 78

SEED. 2006. Insight 31: Review of strategies to address gender inequalities in Scottish schools. Edinburgh: Scottish Executive Education. In: http://www.scotland.gov.uk/Resource/Doc/113682/0027627.pdf [10.1.2013]

Sharpe, S. 1976. Just Like a Girl': How Girls Learn to Be Women. Harmondsworth: Penguin

Simpkins, S. D.; Davis-Kean, P. E. & Eccles, J. S. 2005. Parents' socializing behaviour and children's participation in math, science, and computer out-of-school activities, Applied Developmental Science, 9: 14-30

Skelton, C. & Francis, B. 2009. Feminism and the schooling scandal. London: Routledge

Smith, I. 1996. Gender differentiation: gender differences in academic achievement and self-concept in coeducational and single-sex schools. In: http://alex.edfac.usyd.edu.au/LocalResource/studyl/coed.html [10.1.2013]

Steele, C. M. 1997. A threat in the air: How stereotypes shape intellectual identity and performance, American Psychologist, V 52: 613-629

Steinke, J. 2005. Cultural Representations of Gender and Science Portrayals of Female Scientists and Engineers in Popular Films, Science Communication, 27(1): 27-63

Steinke, J.; Long, M.; Johnson, M. J. & Ghosh, S. 2008. Gender Stereotypes of Scientist Characters in Television Programs Popular Among Middle School-Aged Children. Paper presented to the SCIGroup, AEJMC, Chicago

Suter, C. 2006. Trends in Gender Segregation by Field of Work in Higher Education, in Women in Scientific Careers: Unleashing the potential. Paris: OECD

Tenenbaum, H. R. & Leaper, C. 2003. Parent-child conversations about science: The socialization of gender inequities? Developmental Psychology, 39: 34-47

Updegraff, K. A.; McHale, S. M. & Crouter, A. C. 1996. Gender roles in marriage: What do they mean for girls' and boys' school achievement? Journal of Youth and Adolescence, 25: 73-88

Vázquez, S. 2010. Los dilemas de las jóvenes ingenieras en el sector TIC, in Género y TIC. Presencia, posición y políticas, edited by C. Castaño. Barcelona: Editorial UOC

Vendramin, P.; Valenduc, G.; Guffens, C.; Webster J.; Wagner, I.; Birbaumer, A.; Tolar, M.; Ponzellini, A. & Moreau, M.P. 2003. Widening Women's Work in Information and Communication Technology: Conceptual framework and state of the art. In: http://www.ftu-namur.org/www-ict/ [10.1.2013]

Wajcman, J. 1991. Feminism confronts technology, Cambridge: Polity Press

Weisner, T. S. & Wilson-Mitchell, J. 1990. Nonconventional family lifestyles and sex typing in six year olds, Child Development, 61(6): 1915-1933

Williams, C. L. 1992. The Glass Escalator: Hidden Advantages for Men in the Female Professions, Social Problems, 39: 253-267. In: http://www.jstor.org/stable/3096961 [10.1.2013]

Xie, Y. & Shauman, K. 2003. Women in Science: Career Processes and Outcomes. Cambridge: Harvard University Press

Younger, M. & Warrington, M. 2007. Single-sex classes in co-educational schools, in Genderwatch: still watching, edited by K. Myers. London: Trentham Books

Part 1 – Images of SET, youth and gender

During adolescence young people often struggle with their self-concept. Personal changes in combination with the trial to fulfill societal expectations of being either "feminine" or "masculine" can influence individuals' perception of gender role "adequate" jobs. In their direct environment, in- and outside school, teenagers face different values and norms which are of great importance during this developmental phase. This topic deals with individuals' interconnection of SET images and gender and discusses which attitudes male and female teenagers have towards SET. How does gender influence the interest in these fields? And generally, which imaginations do young girls and boys have about their future?

Why girls stay away from STEM: How the image of science clashes with teenagers' identity

Ursula Kessels

Abstract

In this article, concerning female students' low involvement in SET, the importance of specific *academic interests and dislikes for adolescents' identity development in general* is stressed. I report on the main results of a larger research programme by Bettina Hannover and myself in which we have dealt with the *perceived fit/misfit between adolescents' self-image and the image of science*. We were able to show that students dislike science to the extent that their self-image deviates from the image of the typical student who favours science. If students perceive themselves as being very different from the typical student favouring science, they usually report not liking science, not aiming at a science related career, and not choosing a science related profile at school. We further demonstrate that the *stereotyping of physical science as a masculine domain* makes stronger engagement in physics threatening for female adolescents, because it endangers their developing identity as a "woman-to-be", while avoiding physics will foster and stabilize their gender-identity. We showed that the typical girl liking and excelling in physics is perceived as being very unfeminine, even masculine, and unpopular with boys. As well, other aspects of the typical person who likes science seem highly incompatible with the self-image of most (female) adolescents. In addition, we studied intervention studies that aim at enhancing the perceived fit between girls' self-image and the image of science related subjects.

1   The development of academic interests is related to identity formation as a whole

In our theoretical approach, we closely relate the development of academic interests to children's and adolescents' need to develop and demonstrate their selves. In the following, I summarize the line of argumentation that is more extensively explained in other papers and chapters (Hannover & Kessels 2004; Kessels 2005; Kessels 2007; Kessels & Hannover 2004; Kessels & Hannover 2007; Taconis & Kessels 2009). We have proposed that liking certain school subjects while disliking others serves children and adolescents as a means for developing their identity as a person with specific interests and characteristics (see figure 1).

Figure 1: The development of interest as identity regulation at school

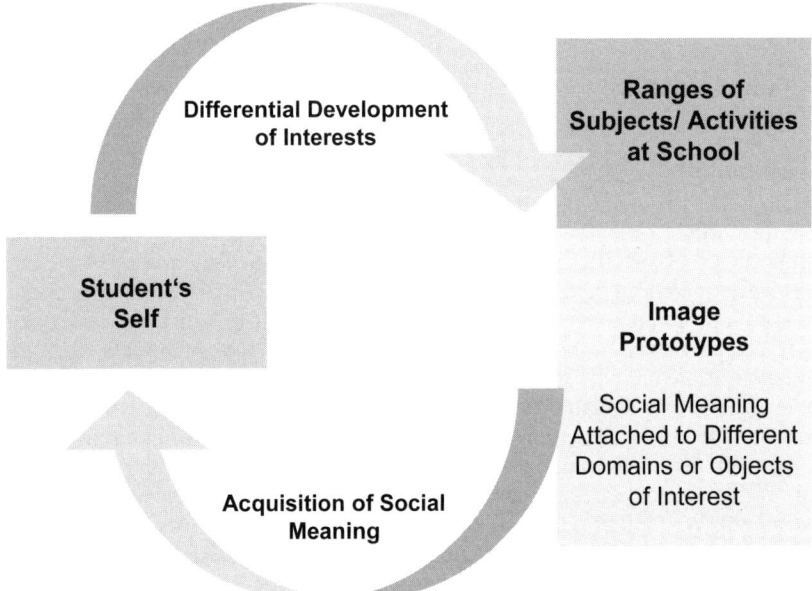

Source: Taken from Kessels & Hannover 2007

In this sense, we understand that a specific profile of interests results from the specific use a student makes of the range of academic subjects, a use that is functional for the development and regulation of his or her identity. Identity or self-concept is defined in our theoretical framework as a person's cognitive representation of him or herself, so that the self-concept comprises all pieces of knowledge that a person has acquired about themselves throughout the life span. The acquisition of a self-concept shapes the personality and the behaviour of an individual (Hannover 1997), with particularly strong impact during the first two decades of life. While individuals move through childhood and adolescence, they are confronted with many changing conditions, biological (e.g., puberty), cognitive (e.g., achievement of different stages of cognitive development), and social (e.g., entering high-school, new peer relationships), in which the child or adolescent continually acquires new self-knowledge, i.e., he or she is continually engaged in defining who and how he or she is. This process of self-definition is an active and self-initiated one (Ruble 1994).

To explain the decline in students' interest towards science related subjects over the course of the school years (e.g. Auriol 2005; Beermann, Heller & Menacher 1992; Eccles [Parsons], Midgley & Adler 1984; Hoffman, Häußler & Lehrke 1998), we suggest that the perceived misfit between most adolescents' self-concept and their image of science is a crucial factor. While the majority of the extant psychological literature on how to explain the lack of students' interest in science subjects and how to augment their participation in these subjects has focused on achievement related variables (such as aptitude, prior knowledge, previous achievement-related experiences, self-concept of ability, attribution of performance outcomes), our approach focuses on the social meaning attached to different domains or objects of interest. We have proposed that by being interested in a certain domain, the student acquires not only knowledge and expertise about this domain, but also the social meaning that is attached to that particular domain. By being interested in a specific domain, its social meaning also becomes part of the student's identity. The social meaning of a school subject is comprised of common assumptions about (a) typical characteristics, contents, and scripts of that subject (we call this the image or stereotype of a school subject), and (b) the characteristics that identify students who have a preference for that subject (prototype of a school subject) (Hannover & Kessels 2004; Kessels & Hannover 2004). As an illustration, if young people share the assumption that someone interested in science is typically a badly dressed and socially isolated "nerd", commitment to this school subject will likely also depend on whether he or she would like to be associated with the specific social meaning attributed to the subjects.

As we have described earlier (see Taconis and Kessels [2009]), researchers in science education have noted similar ways in which being interested in and specialising in science is a process of 'enculturation' of individuals to the specific culture – or even 'subculture' – of science (Aikenhead 2001; Krogh & Thomson 2005; Lyons 2006; Wenger 1999). Even when their focus is not on the psychological mechanism behind this process, they link the specific features of the science culture to adolescents' lack of interest in science. Studies on the perception of science have generally found that science culture is viewed more negatively than positively, a finding that matches and might explain the negative attitudes towards and lack of interest in science encountered in many countries (for overviews, see Driver, Leach, Millar, & Scott 1996; Driver, Newton, & Osborne 2000; Lederman 1992; Osborne, Simon, & Collins 2003; Schreiner 2006; Sjøberg 2002). The specifics of the image of science will be described later in paragraph 1.2.

Schreiner (2006; see also Schreiner & Sjøberg 2007) has also discussed how the refusal to become part of the "subculture of science" is related to specific identity formation processes in our societies. Some parts of her argument are summarised in what follows (cited from Taconis and Kessels [2009]). Schreiner describes how the 'late modern zeitgeist' can be highlighted with the key words of detraditionalisation, cultural liberation, risks, reflexivity, individualisation, and, most importantly and resulting from many of the previous concepts, active identity construction (Schreiner 2006: 36). Such, identity formation has become an individual's personal 'project' requiring deliberate effort and individualistic choices. These choices are related to everyday matters like clothes, taste in music, sports, and beliefs (Giddens 1991), as well as to school and classroom matters. Schreiner concludes that most young people in late modern societies, especially females, choose an identity that is not connected to science and the culture of science because students' perception of the culture of science clashes with their identity projects, which in late modern societies are usually centred around the idea of self-realisation. Although her argumentation is convincing, her work relates – relatively vaguely – different phenomena on a collective level but draws inferences about individual level relationships without having tested them. (This could result in a so-called "ecological fallacy"). However, in our own approach, we have been able to show empirically – on the level of the individual- that a perceived mis-fit between the image of science on the one hand and the self-concept of a student on the other hand is correlated with students' involvement in science.

1.1   Applying the self-to-prototype matching theory to academic choices: The misfit between the image of science and students' identity

In several papers we have emphasised that a crucial factor in students turning away from science is the mismatch between their image of science and most students' sense of identity or self (e.g. Hannover & Kessels 2004; Kessels 2005; Kessels et al. 2006). Applying the self-to-prototype matching theory (Burke & Reitzes 1981; Niedenthal, Cantor & Kihlstrom 1985; Setterlund & Niedenthal 1993) to students' peference for school subjects, we found that the perceived similarity between one's self and the prototype of someone liking or choosing science predicts liking and choosing academic subjects. In this theory, a "prototype" means a cognitive representation of a typical, average 'best example' (Rosch 1973). Niedenthal and colleagues linked the idea of the prototype to choices that people make. They proposed that when making a self-relevant decision, people compare their self-image with the prototype who would choose each of the options in question and eventually select the option with the greatest similarity between self and prototype. People act in this way to conserve or even strengthen their self-image when making self-relevant decisions that actually or symbolically involve entering a specific social context (Niedenthal et al. 1985). In addition, adolescents in particular may search for situations in which they can elicit self-verifying feedback (Ruble 1994; Hannover 1998), as the "defining life task of adolescence [is] the question of discovering who one is, who one belongs with, what one is good at, and where one is going in the future" (Brinthaupt & Lipka 2002: 93), so that the self-to-prototype matching approach seems especially fruitful for understanding teenagers' academic choices and preferences. Importantly, this should be true in general for both male and female students.

We have compared in several studies a description of someone who likes science to students' self-image. In many different samples we have found that the prototypical peer who favours science was highly incompatible with most students' self-image and was seen to possess relatively more negative traits than a prototypical peer who favours languages (Hannover & Kessels 2004; Kessels & Hannover 2002; Taconis & Kessels 2009). Specifically, the science prototype was described as being less physically and socially attractive, less socially competent and integrated, less creative and emotional, although at the same time more intelligent and motivated than peers who prefer languages. In support of the assumptions made in the self-to-prototype matching paradigm, we found that school students' liking of school subjects was stronger the more similar their description of the prototypical peer favouring a particular subject was to their own self-image (Hannover & Kessels 2004). We also showed that choosing science at secondary school was based on the perceived fit between self and the science prototype (Taconis & Kessels 2009).

1.2   The image of SET subjects clashes with most girls' identity

Although the psychological mechanism that links the development of academic interest to students' identity development as a whole is the same for male and female students, the prototype and the image of science do not fit male and female students' identity to the same extent.

We give a brief summary below of findings regarding the image of science with special emphasis on the masculine stereotyping of this domain. The subject science is seen by many students as 'dull, authoritarian, abstract, theoretical, fact-oriented and fact-overloaded, with little room for fantasy, creativity, enjoyment, and curiosity', and 'difficult and hard to understand' (Sjøberg 2002, cited in Schreiner 2006: 57). Studies examining the beliefs of students about science and mathematics have repeatedly revealed that students tend to perceive these subjects as offering fewer opportunities to form and express one's own ideas than arts and languages do (for overviews see Lederman 1992; Osborne, Simons, and Collins 2003). In addition, we found that many teenagers named physics a "boys' subject" (Hannover & Kessels 2002).

Going beyond the extant literature on the image of science, we conducted a study to measure whether and how the image of science is activated and used as a cognitive schema whenever a student encounters a stimulus pertaining to physics. In order to show this, we used an adapted version of the Implicit Association Test (IAT) by Greenwald, McGhee, and Schwartz (1998).

This test operates on the basis of the assumption that if two concepts have come to be associated in memory, they will be connected more quickly when they are encountered. For instance, it should be easier to mentally pair the concept flower with the concept pleasant (e.g. represented by wonderful, rainbow) than the concept flower and the concept unpleasant (e.g. represented by disgust, hatred). In the IAT procedure, respondents have to identify stimulus items that appear on a computer screen and categorise them into one of four superordinate categories. Strength of association between the superordinate categories is measured by comparing the speed of sorting category members in two different conditions: A "stereotype-consistent" condition and "stereotype-inconsistent" condition. The so called IAT effect is calculated by using latency data: The difference in average response latency between the two sorting conditions is calculated and divided by the standard deviation of all latencies for both sorting tasks (Greenwald, Nosek, & Banaji 2003). In our study with students from the 11th grade it emerged that specific dimensions of the image of physics could also be detected on the level of automatic, implicit associations with physics (relative to English): Students more easily associated physics than English with words referring to males (than females), to difficulty (than ease), and to heteronomy (than self-realisation) (Kessels, et al. 2006, Study 1).

That physics is perceived as difficult and as allowing no self-realisation will alienate both boys and girls from that domain, while the association of physics with maleness as opposed to femaleness, the perception of physics as a "boys' subject," will exclusively affect girls.

1.3   The prototype of science: Mismatch with most girls' self-image

We alo found that the prototype of a peer liking or excelling in science fits to a different extent the self-image of girls and boys. In our research we could show that girls who are very interested in or show excellent performance in physics are viewed as being not feminine and unpopular with boys (Kessels 2005). We asked 198 participants from the 8th and 9th grade (age about 15 years) about how girls and boys who liked music or physics would be perceived within their own class (physics being the subject that had been named a "boys' subject" most often and music being the subject that had been named a "girls' subject" most often [Hannover & Kessels 2002]). In this study (Kessels 2005), we measured the ascription of femininity and masculinity to prototypical students who liked physics or music best. The description of prototypes was assessed by participants' rating of the extent to which fifteen feminine and fifteen masculine trait adjectives were characteristics of a) the typical girl favouring physics, b) the typical boy favouring physics, c) the typical girl favouring music, and d) the typical boy favouring music. As expected, the description of prototypical peers on traits of masculinity and femininity varied not only with the sex of the prototype, but also with his or her preferred school subject: girls whose favourite subject was physics were regarded as possessing more masculine traits and less feminine traits than boys whose favourite subject was music (and than girls whose favourite subject was music). In addition, we found in this study that the perceived popularity of boys and girls is linked to their achievement in gender-typed school subjects. Specifically, students believed that male classmates would punish counter-stereotypical performance: Boys were expected to like a girl excelling in physics less than a boy who is the best student in physics and to like a boy doing extremely well in music less than a girl who is the best student in that subject. Female classmates, however, were assumed to simply sanction students who excel in physics: participants expected girls to like someone who is the best student in physics less than someone who is the best student in music – irrespective of the sex of the student. In a second step, we tested how the assumption of being or becoming unpopular with one's classmates when showing a counter-stereotypical achievement profile affects those girls and boys who actually do very well in physics/ music. Results showed that boys did not apply the commonly held stereotypes to themselves. Even with the best grades in music, they did not consider themselves unpopular with other boys. By contrast, girls who did very well in physics did, in fact, indicate they felt rather unpopular with boys (Kessels 2005). Altogether, our findings suggest that the negative prototype of girls liking science contributes to girls' withdrawal from these subjects. They seem to fear – it seems with good reason – being rejected by their male classmates and seen as unfeminine, as it is mainly the reputation of girls that is threatened if they excel at physics.

Following this approach, Rommes, Overbeek, Scholte, Engels and De Kemp (2007) wanted to better understand why girls are underrepresented in computer science. Analysing both individual and focus-group interviews, they also found evidence that professional choices are based on self-to-prototype matching, even when the prototype includes characteristics that are irrelevant for that profession, such as sexual attractiveness.

Taken as a whole, the incompatibility of the science-preferring prototype to female students' self-image is a crucial factor in why female students especially do not wish to specialise in science. Based on this finding, we propose that one reason girls abstain from physics might be that they fear the ascription of any liking for physics will reduce the extent of their being perceived as feminine and popular with the other sex.

In order to demonstrate that girls in fact feel threatened in their identity as a "typical girl" when being perceived as liking or excelling in physics and that they do not engage in physics because they do not want to be perceived as unfeminine, mere correlational data is not sufficient. Consequently, we conducted an experimental study that aimed at demonstrating this mechanism

(Kessels, Holle, Warner & Hannover 2008). In order to show that girls perceive too much closeness to physics as a threat to their femininity, we examined the effects of false positive feedback on boys' and girls' self-presentations in a sample of 135 9th graders (age about 15 years). If our assumptions were correct, girls would react to highly positive feedback concerning their physics' abilities in a compensatory manner: instead of accepting the positive feedback, they should attempt to emphasise their femininity. Boys, however, when confronted with the same highly positive feedback were expected to accept the feedback at face value (instead of engaging in compensatory self-presentations). The experimental treatment consisted of variation in the feedback students received after having worked on a physics test: half of the participants received highly positive feedback, the other half of the participants received average feedback (random assignment). Both types of feedback were embedded in a brief description about the many interesting career options offered today in the realm of science and technology. After having received the feedback, students were given the opportunity to express their interest in reading sex-typed teenager magazine articles (instead of reading articles that referred to the very positive feedback in physics) as a means to demonstrate their femininity or masculinity. Specifically, they were asked to rate how much they would like to read each of 20 different articles of which only the titles were given in a questionnaire (different versions for girls and boys; all titles had been pretested before). Ten titles referred to career options in science and technology (same items in the boys' and the girls' version), while the other ten captured typical sex-typed teenager themes (e.g. "All natural beauty secrets" in the girls' version; "Chatting up girls – what comes next?" in the boys' version). As expected, a significant interaction of the factors gender and feedback emerged: While boys demonstrated relatively more interest in articles dealing with career options in science after having received highly positive feedback on their science ability than after having received only average feedback, this pattern could not be detected with girls. Even when they were given highly positive feedback on their physics ability, they did not choose science related topics but typical girls' topics. Within our theoretical framework, this can be interpreted as girls rejecting the ascribing of high ability in a masculine school subject and attempting to be viewed as typical, feminine girls instead.

To summarize, the negative stereotype of girls in the sciences provides an explanation for why girls in particular do not like these subjects: Girls who succeed in physics consider themselves at risk of not complying with the feminine role in the eyes of others. By turning away from these subjects in school, teenage girls try to avoid being ascribed the negative features of the science-preferring prototype, such as being unfeminine and unpopular with boys.

2 Interventions based on the identity approach

What can we learn from our theoretical framework when focusing on interventions that aim at increasing the number of girls and young women who opt for a science related career? When taking into account that most girls experience a misfit between the image of science and their own self-image, interventions should try to enhance the individually perceived fit between self and science.

Figure 2: Starting points to enhance the fit between students' self-image and image of science

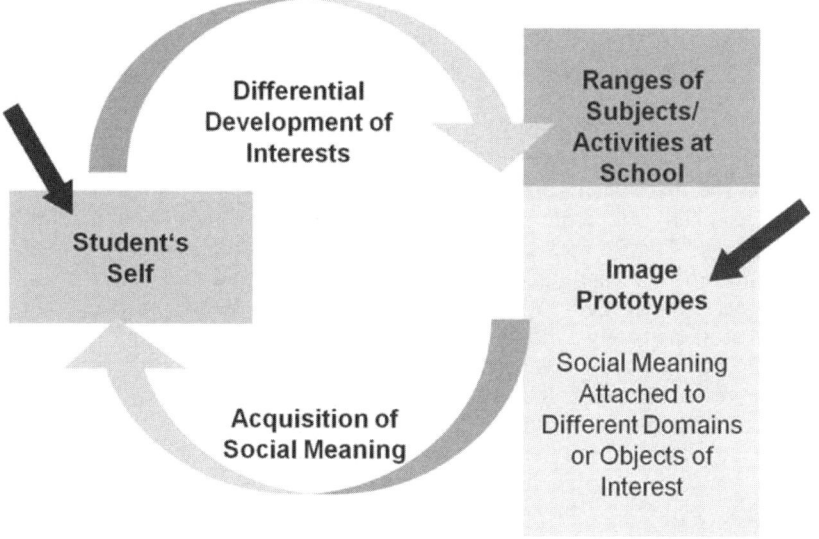

Source: Adapted from Kessels & Hannover 2007

As depicted in figure 2, narrowing the perceived gap between selves and the image of science could be achieved by a) altering something about girls' selves in order to make them a better fit with science or b) altering the image of science in order to make it more similar to girls' selves. In the following, examples from our own research are described, one for each possible route. Since altering the image of science seems at first sight more easily achieved than altering girls' selves, the first section will report on an experimental study in which we tried to alter the gendered connotations of science in order to make it less masculine. In the second section, I will describe how we can deactivate aspects of girls' selves that are highly incompatible with the masculine image of science.

2.1 Altering/Deactivating incompatible aspects of the image of science

Can female role models decrease the strength of the association between maleness and physics? In an experiment originally reported in Kessels and Hannover (2007), female university students were randomly assigned to three experimental conditions. In the experimental treatment condition "female role model", participants read a short person description of a fictitious female physicist. In the treatment condition "male role model", students read the same text, which had been adapted so that it pertained to a male physicist. To ensure that students would build a mental image of the person described, in both conditions they were asked to write down a birthday present they would like to give to the person, were they invited to the birthday party. In the control condition "no role model", they read a text about a landscape in Switzerland. As dependent variable, we measured the automatic associations between physics (as compared with English) and masculine versus feminine words by using an Implicit Association Test (as briefly described in section 1.2). The more easily participants associated physics with masculine words and English with feminine words, as compared to physics with feminine words and English with masculine words, the larger the resulting IAT effect. Results showed that students who read the text about the female physicist produced much smaller IAT effects than students in the two other conditions, so that immediately following the encounter of a female role model, the automatic association between physics and masculinity was diminished. Taken as a whole, this study shows that by altering incompatible aspects of the image of science we can enhance the perceived fit between girls' self-image and science. Our findings suggest that a more balanced representation of male and female role models in science related contexts – such as teachers or role models depicted in the media – would weaken the association between science and the image dimension "gender connotation". Such a change should have a particularly positive impact on girls' implicit attitudes towards the sciences.

2.2 Altering/Deactivating incompatible aspects of girls' self-image

While altering the image of science seems to be an immediate, plausible way to enhance the perceived fit between girls' selves and their perception of science, any interventions that seek to alter aspects of girls' identity might be seen as both more difficult to achieve and less acceptable.

Based, however, on the presupposition that an individual's identity or self is a changing and flexible construct rather than a static one, this approach might make more sense than is immediately apparent. Research from social cognition has pointed out that individuals' identity varies according to the social context in which the individual is found (e.g. Hannover 1997; Linville & Carlston 1994). At any given moment, some but not all parts of a person's extensive self-knowledge are activated. For instance, in a situation where gender is psychologically salient the individual may retrieve knowledge related to gender easily, while on a different occasion this self-knowledge might be difficult to access. Markus and Kunda (1986) called this situationally accessible self-knowledge the 'working self'.

Linking this research to girls' involvement in SET subjects, we have suggested (Kessels & Hannover 2008) that that the more salient gender-related self-knowledge is in a girl's working self during physics lessons, the worse her physics related self-concept of ability will be. Because physics is seen as a non-feminine subject (see above), girls will perceive a larger misfit between themselves and physics in a context in which their identity as female is activated.

What kind of contextual factors influence whether or not a person's gender related self-knowledge is activated and can any of these factors be manipulated in science lessons? One factor influencing the salience of a person's gender is the sex composition of the group the person is in. Studies found that gender is more salient in mixed-sex groups than in single-sex groups (e.g. Cota & Dion 1986; McGuire & Padawer-Singer 1976). If all group members are of the same sex, gender is not a useful category for describing or differentiating any of them, so that in a single-sex group, gender-related self-knowledge should be less accessible than in a mixed sex group. Consequently, girls' self-concept regarding ability in physics should be more positive in single-sex groups, because for girls, this lower accessibility should lead to an increased fit between their working selves and the subject of physics, resulting in a better self-concept of ability in physics. We tested this assumption in a quasi-experimental study (Kessels & Hannover 2008) with students from four state coeducational schools who were randomly assigned to single-sex or mixed-sex classes in physics throughout their first year of physics instruction (8th grade). In an experiment measuring response latencies during these physics lessons, our hypothesis that gender-related self-knowledge was more accessible in mixed-sex classes than in single-sex classes was supported: Adolescents in mixed-sex classes were faster at responding to gender-typical traits than adolescents in single-sex classes. And while girls in mixed-sex classes were responding relatively faster to feminine traits than to masculine traits, this difference was attenuated in single-sex classes. We had proposed that these findings on the reduced accessibility of gender-related self-knowledge in single-sex classes might explain why girls develop a higher self-concept of ability in physics in these groups. We expected that girls' self-concept of ability in physics would depend on the extent to which their 'incongruous' feminine self-knowledge and their 'congruous' masculine self-knowledge were accessible during physics lessons (the pattern more likely to be encountered in girls-only classes). Our study showed that the less accessible feminine self-knowledge and the more accessible masculine self-knowledge were during physics lessons, the better was girls' self-concept of ability in physics (Kessels & Hannover 2008).

Taken as a whole, this study shows how the deactivation of gender-related self-knowledge in the school context can enhance the perceived fit between girls' self-image and science.

References

Aikenhead, G. 2001. Student's ease in crossing cultural borders into school science. Science Education, 85, 180-188

Auriol, L. 2005. Are there less students choosing science and engineering studies? Presentation at the International OECD Conference 'Declining Student Enrolment in Science & Technology', Amsterdam; Netherlands, 14.-15.11.2005

Beerman, L.; Heller, K. A. & Menacher, P. 1992. Mathe: nichts für Mädchen? [Math: not for girls?]. Bern: Huber

Brinthaupt, T. & Lipka, R. 2002. Understanding early adolescent self and identity: An introduction. In T. Brinthaupt & R. Lipka (Hrsg.). Understanding early adolescent self and identity. Applications and interventions. Albany: State University of New York Press: 1-21

Burke, P. J., & Reitzes, D. C. 1981. The link between identity and role performance. Social Psychology Quarterly, 44: 83-92

Cota, A. A. & Dion, K. L. 1986. Salience of gender and sex composition of ad hoc groups: An experimental test of distinctiveness theory. Journal of Personality and Social Psychology, 50: 770-776

Driver, R.; Leach, J.; Millar, R. & Scott, P. 1996. Young people's image of science. Buckingham, UK: Open University Press

Driver, R.; Newton, P. & Osborne, J. 2000. Establishing the norms of scientific argumentation in classrooms. Science Education, 84: 287-312

Eccles (Parsons), J. S.; Midgley, C. & Adler, T. 1984. Grade-related changes in the school environment: Effects on achievement motivation. In: Nicholls, J. G. (ed.), The Development of Achievement Motivation, JAI Press, Greenwich, CT: 283-331

Greenwald, A. G.; McGhee, D. E.; & Schwartz, J. L. K. 1998. Measuring individual differences in implicit cognition: The implicit association test. Journal of Personality and Social Psychology, 74: 1464-1481

Greenwald, A. G.; Nosek, B. A. & Banaji, M. R. 2003. Understanding and using the implicit association test: I. An improved scoring algorithm. Journal of Personality and Social Psychology, 85: 197-216

Hannover, B. 1997. Das dynamische Selbst. Zur Kontextabhängigkeit selbstbezogenen Wissens. [The dynamic self. The context-dependency of self-related knowledge.] Bern, Switzerland: Huber

Hannover, B. 1998. The development of self-concept and interests. In: L. Hoffmann; A. Krapp; K. A. Renninger & J. Baumert (Eds.), Interest and learning. Kiel: IPN: 105-125

Hannover, B. & Kessels, U. 2002. Challenge the science-stereotype! Der Einfluss von Freizeit-Technikkursen auf das Naturwissenschaften-Stereotyp von Schülerinnen und Schülern [Challenge the science-stereotype! The influence of optional science courses on student's sciences stereotype]. Zeitschrift für Pädagogik, 43: 341-358

Hannover, B. & Kessels, U. 2004. Self-to-prototype matching as a strategy for making academic choices. Why German high school students do not like math and science. Learning and Instruction, 14 (1): 51-67

Hoffmann, L.; Häußler, P. & Lehrke, M. 1998. Die IPN-Interessenstudie Physik [The physics interest study of the Institute for Science Education]. Kiel: IPN

Hoffmann, L.; Häußler, P. & Peters-Haft, S. 1997. An den Interessen von Mädchen und Jungen orientierter Physikunterricht. Ergebnisse eines BLK-Modellversuchs. Kiel: IPN

Kessels, U. 2005. Fitting into the stereotype: How gender-stereotyped perceptions of prototypic peers relate to liking for school subjects. European Journal of Psychology of Education, 20 (3): 309-323

Kessels, U. 2007. Identitätskongruente Nutzung des schulischen Angebots. [Fitting academic interests to identity] Habilitation: Freie Universität Berlin

Kessels, U. & Hannover, B. 2004. Entwicklung schulischer Interessen als Identitätsregulation. [The development of academic interests as identity regulation] In: J. Doll & M. Prenzel (Hrsg.). Bildungsqualität von Schule: Lehrerprofessionalisierung, Unterrichtsentwicklung und Schülerförderung als Strategien der Qualitätsverbesserung. Münster: Waxmann: 398-412

Kessels, U. & Hannover, B. 2007. How the image of math and science affects the development of academic interests. In: M. Prenzel (Ed.). Studies on the educational quality of schools. The final report on the DFG Priority Programme. Münster: Waxmann: 283-297

Kessels, U. & Hannover, B. 2008. When being a girl matters less. Accessibility of gender-related self-knowledge in single-sex and coeducational classes. British Journal of Educational Psychology, 78 (2): 273-289

Kessels, U.; Warner, L. M.; Holle, J. & Hannover, B. 2008. Identitätsbedrohung durch positives Leistungsfeedback. Die Erledigung von Entwicklungsaufgaben im Konflikt mit schulischem Engagement. [Threat to identity through positive feedback about academic performance. Developmental tasks clashing with academic involvement] Zeitschrift für Entwicklungspsychologie und Pädagogische Psychologie, 40: 22-31

Krogh, L. B. & Thomsen, P. V. 2005. Studying students' attitudes towards science from a cultural perspective but with a quantitative methodology: Border crossing into the physics classroom. International Journal of Science Education, 27: 281–302

Lederman, N. G. 1992. Students' and teachers' conceptions of the nature of science: A review of the research. Journal of Research in Science Teaching, 26(9): 771-783

Linville, P. W. & Carlston, D. E. 1994. Social cognition of the self. In: P. G. Devine, D. L. Hamilton & T. M. Ostrom (Hrsg.), Social cognition: Impact on social psychology. San Diego: Academic Press: 144-193

Lyons, T. 2006. The puzzle of falling enrolments in physics and chemistry courses: Putting some pieces together. Journal Research in Science Education, 36(3): 285-311

Markus, H. & Kunda, Z. 1986. Stability and malleability of the self-concept. Journal of Personality and Social Psychology, 51: 858-866

McGuire, W. J. & Padawer-Singer, A. 1976. Trait salience in the spontaneous self-concept. Journal of Personality and Social Psychology, 33: 743-754

Niedenthal, P. M.; Cantor, N. & Kihlstrom, J. F. 1985. Prototype matching: A strategy for social decision making. Journal of Personality and Social Psycholgy, 48(3): 575-584

Osborne, J. F.; Simon, S. & Collins, S. (2003). Attitudes towards science: A review of the literature and its implications. International Journal of Science Education, 25(9): 1049-1079

Rommes, E.; Overbeek, G.; Scholte, R.; Engels, R. & De Kemp, R. 2007. "I'm not interested in computers". Gender-based occupational choices of adolescents. Information, Communication & Society, 10: 299-319

Rosch, E. 1973. Natural categories. Cognitive Psychologist, 4: 328-350

Setterlund, M. B. & Niedenthal, P. M. 1993. "Who am I? Why am I here?" Self-esteem, self-clarity, and prototype matching. Journal of Personality and Social Psychology, 65: 769-780

Ruble, D. N. 1994. A phase model of transitions: Cognitive and motivational consequences. In: M. P. Zanna (Ed.), Advances in experimental social psychology. San Diego, CA: Academic Press, Vol. 26: 163-214

Schreiner, C. 2006. Exploring a ROSE-garden: Norwegian youth's orientations towards science – Seen as signs of late modern identities. Doctoral dissertation, Department of Teacher Education and School Development, University of Oslo, Norway

Schreiner, C., & Sjøberg, S. 2007. Science education and youth's identity construction – Two incompatible projects? In: D. Corrigan; J. Dillon & R. Gunstone (Eds.), The re-emergence of values in the science curriculum. Rotterdam, The Netherlands: Sense Publishers: 231-249

Sjøberg, S. 2002. Science and technology education in Europe. Current challenges and possible solutions. In: E.W. Jenkins (Ed.), Innovations in science and technology education (Vol. VIII). Paris: United Nations Educational, Scientific and Cultural Organization: 201-229

Taconis, R. & Kessels, U. 2009. How choosing science depends on students' individual fit to the "science culture". International Journal of Science Education, 31: 1115-1132

Wenger, E. 1999. Communities of practice: Learning, meaning, and identity. Cambridge, UK: Cambridge University Press

Images of an ideal engineer and self-image – differences between male and female engineering and non-engineering students

Martina Endepohls-Ulpe and Judith Ebach

Abstract

The presented study examines gender differences between male and female technology students, as well as differences between students of technology and non-technology students concerning their self-image and their images of an ideal engineer. The sample of this questionnaire study consisted of 179 non-engineering students and 141 engineering students. The results revealed that the image of an ideal engineer in general is closely related to the male stereotype. Differences in the self-images of male and female engineering and non-engineering students reflect these stereotypes. The self-images of engineering students appear to be closer to their own image of an ideal engineer than those of their peers in non-technological studies. Furthermore, their images of an ideal engineer seem to be more modern or progressive.

Introduction

Several studies conducted by the European Union (e.g. Eurostat 2004; Implementation of "education & training 2010" work programme; Womeng 2002-2005) demonstrate that women and girls are continuously dramatically underrepresented in science and technological education, areas and jobs. Young women consider artistic and health professions as top choices while young men choose science and information technology related careers. These gender specific choices can not only be observed in Europe but also the USA and Canada (Lupart, Cannon & Telfer 2004).

The study presented is part of a European research project (UPDATE, see also Metternich-Neuhäuser & Krummacher in this volume) that aims at improving science and technology teaching in Europe, especially for girls. UPDATE stands for: Understanding and Providing a Developmental Approach to Technology Education[16]. The innovative aspect of the project is that the approach includes a strong focus on early childhood and primary education, phases in which attitudes and behaviours concerning technology

---

16 UPDATE is part of the 6th framework programme of the European Union, period of sponsorship 2007-2009.

are often formed. The objective of the study is also to develop innovative pedagogical practices and learning environments that encourage girls and young women to choose courses of study and professions in the fields of natural science and technology. This is done in the 11 participating countries by exploring barriers and motivating factors for girls to take up courses of study from science and technology.

Figure 1: Eccles Model of Achievement related choices in education and career decision making

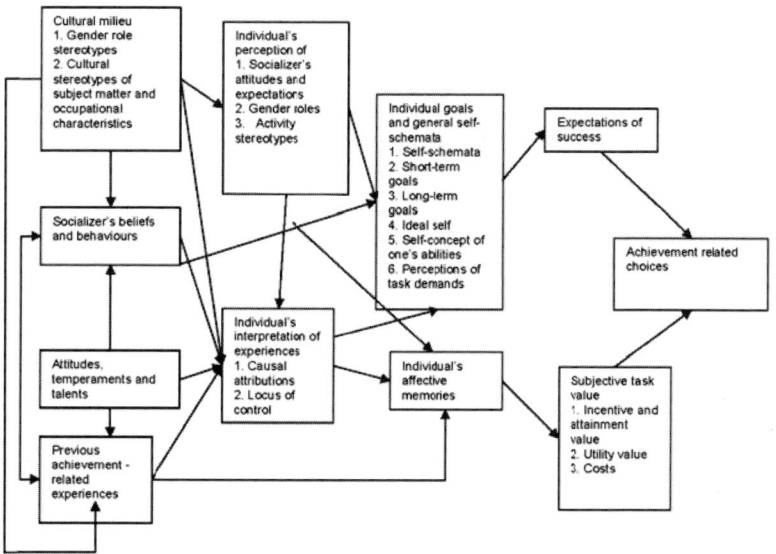

Source: Eccles, 1985, cit. from Lupart et al. 2004

A well-known model for the explanation of gender differences concerning occupational and educational choices was presented by Eccles (1985; 1994). This model points out that decisions concerning apprenticeships or careers base on a complex network of variables. A person's abilities or skills are only one variable among others. Some of the factors which Eccles and her colleagues identified to be influencing achievement-related behaviour and occupational choices are: A person's expectations of success, his or her sense of personal efficacy, the subjective value attached to each option available, gender roles and other identity-related variables (Deaux & Lafrance 1998). The choices that men and women make are influenced by a calculation of the fit between a particular option and a person's basic goals, motivations and self-definitions, as well as by the balance between the attainment value and the perceived costs of this option. Stereotype views of certain professions also play an important role in this process (see figure 1).

Hannover (2004) presents an approach derived from the cognitive psychological theory of gender role development which she names: "Development as regulation of identity". In order to explain gender related behaviour, this approach focuses on the activity of the individual (child or adolescent) who constructs his or her identity as a girl or boy. This is equally true for young children as for adolescents in puberty, a time when gender as a category for identity formation gets a greater relevance for a second time. At this age, young people look for gender differences in their social environment and try to integrate them in their self-concept to verify their gender identity. Choosing a career or even only some courses in secondary school in a field that is strongly related to the male stereotype hence may be a difficulty for girls in their process of gender identity formation and the probability that girls avoid fields that are male dominated rises exactly at a point where the course for the later field of occupational choices is set.

Hannover herself (Hannover & Kessels 2004) delivers empirical support for the position which is held in the two models presented above, namely that adolescents try to choose activities and (achievement related) choices that are in accordance with their self image.

In a study with 104 students from the 9th and 10th grade, the perceived consistencies between the students' self-image and the prototype of a person who is engaged in a specific field of activities influenced the students' tendency of engagement in this field. The students judged the prototype of a student who favours humanities (German and English language) generally more positively than a student who does not like these subjects, whereas for science (maths and physics) the prototypical students who do not like this field were perceived more positively. The better the match between self and favourite subject prototype, the stronger was the own subject preference. This was true for both girls and boys.

According to the presented models and empirical results, the hypothesis suggests that if the prototypical image of a person, who is engaged in a certain occupational field, e.g. engineering, is strongly gender-stereotyped, this circumstance may bring about gender differences in occupational choices. However, images of persons in certain occupations as well as the grade of conformity of the self-image with gender stereotypes may differ between individuals and these differences also may cause differences in occupational choices.

Engineering has been traditionally defined as a male domain. Paulitz (2009) demonstrates that different social constructions of the "classical engineer" have existed over the last two centuries (for the German tradition see

Zachmann 2002), but they all somehow were defined in a way that excluded women.

For the development of new strategies to encourage girls to choose professions from this field, it would be interesting to see how the self-images of young women who choose courses of study in an engineering profession are related to their image of a successful engineer and if their self-images or their images of an engineer are different from those of students who choose courses from other fields, e.g. teaching. Hence, the presented study explores the images of a person who works successfully in the field of engineering held by both, male and female students of engineering and non-engineering courses of study, and compares them to their self-images.

Method

Participants

The total sample of the study consisted of 320 students:
- 179 non-engineering students (49 male (27.4%) and 130 female (72.6%)), (University of Koblenz-Landau)
- 141 engineering students (79 male (56.0%) and 62 female (44.0%)) (Several universities of applied sciences in Rhineland-Palatinate, federal state of Germany).

Table 1: Course of study of non-engineering students, sex

Subject	Male N (%)		Female N (%)		Total N (%)	
Educational science	4	(8.2)	10	(7.7)	14	(7.8)
Teaching in primary school	19	(38.8)	78	(60.0)	97	(54.2)
Teaching in secondary school	21	(42.9)	36	(27.7)	57	(31.8)
Other	5	(10.2)	6	(4.6)	12	(6.7)
Total	49	(100.0)	130	(100.0)	179	(100.0)

Source: Own table

The age of the students differed between 19 and 48 (mean 23.39, median 22), sub-samples (non-engineering vs. engineering; male vs. female) didn't differ significantly in age.

The distribution of non-engineering students on different courses of study (teaching, educational science) and engineering students on courses with different main subjects is shown in table 1 and table 2.

Table 2: Course of study of engineering students, sex

Subject	Male N	(%)	Female N	(%)	Total N	(%)
Medical technology	17	(21.5)	11	(17.7)	28	(19.9)
Laser technology	23	(29.1)	12	(19.4)	35	(24.8)
Mechanical engineering	14	(17.7)	10	(16.1)	24	(17.0)
Food technology	4	(5.1)	7	(11.3)	11	(7.8)
Environment protection	1	(1.3)	9	(14.5)	10	(7.1)
Computational sciences	3	(3.8)	4	(6.5)	7	(5.0)
Applied informatics	6	(7.6)	0	(0.0)	6	(4.3)
Computing engineering	4	(5.1)	1	(1.6)	5	(3.5)
Bio-technology	1	(1.3)	0	(0.0)	1	(0.7)
Electrical engineering	4	(5.1)	1	(1.6)	5	(3.5)
Measurement technology	2	(2.5)	2	(3.2)	4	(2.8)
Other	0	(0.0)	5	(8.1)	5	(3.5)
Total	79	(100.0)	62	(100.0)	141	(100)

Source: Own table

Instruments

The administered questionnaire consists of 17 4-point Likert-Items (1 = "do not agree at all" to 4 = "totally agree") and is an extended version of a scale originally developed for the WOMENG study (5th FP of the EU commission, funding period 2002-2005) (Pourrat et. al 2005). In the WOMENG study, the scale was used to describe a male engineer in general, while in the presented study, the students filled in the scale first of all to describe their image of an ideal engineer, who works successfully in his/her job, and secondly to describe their self-image. The scales were part of a larger questionnaire which aimed at the identification of motivating factors and barriers to study engineering (Ebach, Endepohls-Ulpe & von Zabern 2009). In this questionnaire, the two scales were placed at very different positions and the items appeared in a different order to avoid answering tendencies.

Statistical analysis

To test the hypothesis that the self-images of engineering students are more similar to their ideal images of an engineer than the self-images of non-engineering students, for each of the 17 characteristics of the questionnaire the absolute difference between the self-description and the description of an ideal engineer was calculated, summed up and divided by 17.

An analysis of variance was performed with values of mean difference as dependent variable and gender and engineering/non-engineering course of study as main factors.

The self-descriptions on the 17Item scale were subjected to a factor analysis to reduce the number of dimensions. For the resulting five dimensions mean scores of substantially loading Items were calculated. This was done as well for the self-descriptions as for the descriptions of an ideal engineer.

Significant differences between the subgroups on the mean scores of the dimensions were tested by analysis of variance with gender and course of study as between subject factors and scores for the five dimensions of image of an engineer and self-image as within subjects repeated measurement factor.

Results

The analysis of variance for the values of mean differences between self-images and image of an ideal engineer revealed a significant difference between engineering and non-engineering students (F = 52,734; p < .001). The self-descriptions of the engineering students were significantly closer to their images of an ideal engineer (mean difference: M = .61, SD = .27; male M = .62, SD = .23; female M = .61, SD = .20) than the self-descriptions of the non-engineering students (mean difference .87; SD = 27). Female non-engineering students seemed to be the subgroup with the greatest difference between self-image and image of an engineer (M = .90; SD = .26; male M = .79, SD = .29)), though the significance for the interaction between course of study and gender can only be interpreted as a weak tendency (p = .06).

Absolute differences between self-image and image of an engineer of course do not give any information about the nature of the two constructs in the subsamples. Furthermore, the results presented above could be a consequence of differences between non-engineering students and engineering students in their self-images or in their images of an ideal engineer or both. To find out the divergences in content, we compared the four subsamples with respect to mean values on the five factors of the self-image scale and the corresponding values of the descriptions of an ideal engineer.

The five factors, which explained 61.25% of the total variance, were named as:

I Social competence and managerial skills
II Abstract and analytical thinking
III Technical competence and machine orientation
IV Creativity and innovativeness
V Ambition and industriousness

For factor I, "social competence and managerial skills", there was a significant main effect (F = 59.35; p < .001): The students in general described themselves as socially more competent (M = 3.2, SD = .48) than they described an ideal engineer (M = 2.9, SD = .55). There was also a significant interaction effect for Factor I X course of study (p < .001). Engineering students rated their self-image concerning social competences and managerial skills as high (M = 3.11, SD = .51) as they considered the social and managerial skills of an ideal engineer (M = 2.99, SD = .55), while non-engineering students rated their own social competences as very high (M = 3.2, SD = .45) but the social competences and managerial skills needed by an engineer as significantly lower (M = 2.83, SD = .54). This was true for both male and female students.

Table 3: Loadings of the questionnaire items on the 5 main factors

	Factors				
	1	2	3	4	5
Being content with private life	.722				
Being able to balance private life and work	.683				
Having social competence	.643				
Being self confident	.604		.465		
Having leading qualities	.537		.385		
Being team-oriented	.489			.333	.364
Thinking in a complex way		.702	.421		
Thinking straight		.702			
Being able to think logically		.698	.310		
Thinking abstractly and analytically		.641	.371		
Being concrete and pragmatic		.440	.356		
Having technical competences			.811		
Being machine-oriented			.777		
Being innovative				.810	
Being creative and innovative				.783	
Being hard working					.884
Being ambitious	.334				.728

Source: Own table

For the second factor, "abstract and analytical thinking", there was also a general significant difference between self-images and images of an ideal engineer (F = 11.85; p < .01): Students rated the capacity of abstract and logical thinking needed for being a successful engineer higher (M = 3.17, SD = .51) than they did for their self-image (M = 3.0, SD = 54). The analysis also revealed significant interaction effects between analytic thinking X course of study (F = 8.909; p < .01) and thinking capacity X gender (F = 10.192; p < .01). Students of engineering courses portrayed their competences in abstract and analytical thinking as high (M = 3.08, SD = .49) and did the same for their images of an ideal engineer (M = 3.08, SD = .51), while non-engineering students portrayed their own intellectual capacities in abstract and logical thinking as high (M = 2.93, SD = .56) but their images of an ideal engineer showed a significantly higher value for this factor, higher than in their self-images and higher than the values of the engineering students (M = 3,22, SD = .49). Male students' self-images concerning their intellectual capacities in general were significantly more positive (M = 3.1, SD = .53) than female students' self-images (M = 2.89, SD = .56), a difference which did not appear in their descriptions of an ideal engineer.

With respect to Factor III, technical competence and machine orientation, a significant main effect shows that the self-images of the students concerning these features (M = 2.3, SD = .81) are significantly lower (F = 141.795, p < .001) than for their images of an ideal engineer (M = 3.05, SD = .60). The significant interaction effect between course of study and Factor III (F = 82.828, p < .001)) reveals that this effect is mainly caused by the great differences between self-images (M = 1.98, SD =.77) and images of an engineer (M = 3.15, SD=.59) perceived by the non-engineering students. Non-engineering students see themselves as not very competent in this field. For engineering students, self-image (M = 2.77, SD = .65) and image of an ideal engineer (M = 2.91, SD = .59) are not so far apart. There is also a significant interaction effect between Factor III X gender (F = 7.97, p < .01). Male students generally score higher on this factor concerning their self-images (M = 2.6, SD = .78) than female students do (M = 2.09, SD = .76), but there is no difference between male (M = 3.0, SD = .60) and female students in general (M = 3.04, SD = .60) with respect to their image of an ideal engineer concerning technical competences and machine orientation. There is a weak tendency for female engineering students (M = 2.78; SD = .57) to judge this feature even as less important than male engineering students do (M = 3.02; SD = .59) (F = 3.26, p = 0 .07).

There are also two significant effects for the between subject factors gender (F = 15.891, p < .001) and course of study (F = 12.786, p < .001) for Factor III. Male students score significantly higher than female students and engineering students score higher than non-engineering students.

For Factor IV, creativity and innovativeness, there was only a significant main effect (F = 72.856, p < .001). Students in general rated their own creativity and innovativeness as high (M = 2.94, SD = .66), but still significantly lower than they considered it to be appropriate for an ideal engineer (M = 3.31, SD = .61).

With respect to Factor V, ambition and industriousness, a weak main effect appeared (F = 5.859, p < .05): Students' self-images concerning this factor were positive (M = 3.06, SD = .74) but their images of an ideal engineer were even slightly more positive (M = 3.16, SD = .65). There was also a significant interaction effect between Factor V X gender (F = 11.325, p < .01): Male and female students differed in their self-image concerning this factor, (male M = 2.88, SD=.75; female M = 3.18, SD = .72), but not in their descriptions of an ideal engineer (male M = 3.12, SD = .65; female M = 3.14, SD = 65).

Summary and discussion

First of all, it can be stated that in accordance with the models of gender typed occupational choices presented above, the self-images of non-engineering students differed considerably more from their images of an ideal engineer than the self-images of engineering students did. Particularly the self-descriptions of the female non-engineering students showed the greatest difference to their images of an ideal engineer. This outcome can be traced back to the fact that their image of an ideal engineer is very close to the male stereotype and will be discussed below when having a closer look at the results in detail.

In general, the congruencies and differences between self-image and image of an ideal engineer in the two subgroups can have two reasons: Differences between the self-images of engineering and non-engineering students or differences in the images of an ideal engineer held by the two groups of students.

Concerning their self-images, the engineering students described themselves as socially competent, well able to think logically and abstractly, moderately machine-oriented and technically competent, creative and innovative and also as ambitious and industrious. Non-engineering students' self-descriptions were even more positive concerning social competence and equally positive for thinking capacity, but they didn't judge themselves as very competent in the field of technical competences and described themselves as absolutely not machine-oriented. Like engineering students, non-engineering students considered themselves as being creative and innovative and rated their ambition and industriousness as high. Hence, main differences in self-images between the two subgroups were the non-engineering students' more positive self-image concerning social competences and their significantly lower self-images concerning their technical competences and machine orientation.

Looking at the images of an ideal engineer held by the two groups and the parallels and differences to their self-images, some remarkable differences can be observed: While engineering students rated the social competences needed by an ideal engineer as high and similar to their self-images with respect to this factor, non-engineering students rated their own social competences as very high and significantly higher than they considered these competences to be necessary to work successfully as an engineer. With respect to thinking abilities, there are also differences in non-engineering students' self-image and their image of an ideal engineer. Their self-images with respect to this feature are positive and do not differ from those of the non-engineering students, but the amount of analytical and logical thinking capacity they regard to be necessary for an ideal engineer is much higher than the abilities they attribute to themselves.

With respect to factor III, technical competences and machine orientation, the discrepancy between self-image and image of an ideal engineer perceived by the non-engineering students is enormous. This is mainly caused by their low self-image concerning this factor, but they also judge these features as more important for being a successful engineer than the engineering students do.

Creativity and innovativeness seem to be very important for being an ideal engineer in the eyes of all students and they assess their own abilities in this field as high but not as high as they are needed by an ideal engineer. The same discrepancy between self-image and image of an ideal engineer can be stated for factor V, ambition and industriousness: The students in general rated themselves lower on these features than they regarded them to be necessary for an ideal engineer.

Are there any indicators for the importance of gender stereotypes in the process of choosing courses of study in the presented data? Looking at the five dimensions included in the administered questionnaire, a typical male person might be portrayed as socially moderately competent, well able to think logically and analytically and having technical competences. Males are supposed to be more innovative and females to be more creative. And normally males are portrayed as less industrious than females.

There was no difference for male and female students in general concerning their self-descriptions with respect to social competences. But in general, male students as a group saw themselves as being better in logical an analytical thinking. This higher appraisal of own cognitive abilities for males is a well-known phenomenon especially with respect to male stereotyped tasks. This tendency can also be observed for technical competences and machine orientation: Male students rate themselves higher on these features than female students do. The higher self-ratings for female students for factor V, industriousness and ambition are also conform to gender stereotypes.

With respect to their self-images, non-engineering students differ from this pattern with regard to their social competences, which they consider to be very high, and their technical competences, which they regard as being very low; this is especially true for the female non-engineering students.

With respect to their images of an ideal engineer, non-engineering students consider the social competences needed by an ideal engineer to be only moderately high, abilities in analytical and logical thinking as well as the technical competences and machine orientation however to be very high. Hence, the images of an ideal engineer held by non-engineering students are indeed very close to the male stereotype, while their self-images are not in some respect. Engineering students' images of an ideal engineer appear to be more realistic and also more apart from the male stereotype: They regard social competences to be more important and analytical and logical thinking as well as technical competences and machine orientation as to be important but not to such an extreme extent. This is especially true for female engineering students, who tend to judge the technical competences and machine orientation needed by an ideal engineer even lower as their male peers do.

Finally, it can be summarized that on the one hand the images of a successful engineer held by students who did not choose a career in this field, are indeed close to the male stereotype, whilst their self-images are not, especially with respect to technical competences. On the other hand the images of an ideal engineer held by engineering students differ from this male stereotyped pattern. They judge social competences as being more important and they see the amount of nonverbal intelligence and technical competences needed to work successfully in this field on a more realistic level. Especially female engineering students differ from their female non-engineering peers with respect to their self-assessed technical competences. They judge their own competences in these fields as higher and, as a consequence, closer to their own image of an ideal engineer than female non-engineering students do.

Conclusion

When aiming at motivating and encouraging especially young women to choose apprenticeships or courses of study from the field of technology, two main strategies can be derived from these results:

On the one hand there is an apparent need for strengthening girls' self-confidence concerning technology. Results of empirical research on gender role development and studies on interest and self-confidence of young girls with respect to technical activities show that this has to take place in kindergarten and primary school (Endepohls-Ulpe, Stahl-von Zabern & Ebach 2009). On the other hand more effort should be put in communicating a realistic and modern image of an engineer to young people in secondary school age, when they begin to make occupational related choices. This can of course be done by presenting a modern image of engineering that is compatible with being feminine in the media. Another option can be seen in mentoring programmes as e.g. the Ada-Lovelace-Project (Ebach 2006) or Cyber-Mentor (Schimke, Stöger & Ziegler 2007); these are programmes that try to spark girls' interest for apprenticeships and courses of study from the field of STEM by bringing them into contact with elder students of these fields or with women who are already successfully working in a field of science, technology, engineering or mathematics.

References

Deaux, Kay & Lafrance, Marianne. 1998. Gender. In: D. T. Gilbert, S. T. Fiske & G. Lindzey, The Handbook of Social Psychology. Boston et al: McGraw-Hill: 788-827

Eccles, Jacquelynne S. 1985. Why doesn't Jane run? Sex differences in educational and occupational patterns. In: F.D. Horowitz & M. O'Brien (Eds.). The gifted and talented: developmental perspectives. Washington, DC: American Psychological Association: 251-295

Eccles, Jacquelynne S. 1994. Understanding women's educational and occupational choices: Applying the Eccles et al. model of achievement related choices. Psychology of Women Quarterly 18: 585-609

Ebach, Judith. 2006. Die Erweiterung der beruflichen Orientierung von Schülerinnen am Beispiel des Ada-Lovelace-Projekts. In: M. Endepohls-Ulpe & A. Jesse, Familie und Beruf – weibliche Lebensperspektiven im Wandel. Frankfurt: Peter Lang: 67-82

Ebach, Judith; Endepohls-Ulpe, Martina & von Zabern, Janine. 2009. Motivating Factors and Barriers to Study Enineering – Gender Differences and Possible Consequences. In: VDI – the Association of German Engineers (Ed.). Gender and Diversity in Engineering and Science. First European Conference. Düsseldorf: 215-230

Endepohls-Ulpe, Martina; Stahl-von Zabern, Janine & Ebach, Judith. 2009. Suggestions for an Improvement of Technology Education in Primary Schools – a Gender Perspective. Paper presented at the ECER conference "Theory and Evidence in European Educational Research", Vienna, Austria, 25-30 Sept. 2009

Hannover, Bettina. 2004. Gender revisited: Konsequenzen aus PISA für die Geschlechterforschung. Zeitschrift für Erziehungswissenschaften, 7, Beiheft 3: 81-99

Hannover, Bettina & Kessels, Ursula. (2004). Self-to-prototype matching as a strategy for making academic choices. Why high school students do not like math and science. Learning and Instruction 14: 51-67

Lupart, Judy L.; Cannon, Elizabeth & Telfer, Jo Anne. 2004. Gender differences in adolescent academic achievement, interests, values and life-role expectations. High Abilitiy Studies, 15, 1: 25-42

Paulitz, Tanja. 2009. Technology and Masculinity – reconsidered – New perspectives on an apparently simple question. In: VDI – the Association of German Engineers. Gender and Diversity in Engineering and Science. Düsseldorf: VDI

Pourrat Yvonne. (ed.) 2005. Methodological Tools for Research in Gender and Technology, Paris, ECEPIE editions

Schimke, Diana; Stöger, Heidrun & Ziegler, Albert. 2007. CyberMentor: E-Mentoring to strengthen interest and participation of girls in STEM. Science in school, 5: 41-44

Zachmann, Karin. 2002. Engendering engineering & engineering gender: German engineers and their professional identities from the 1860s to the 1960s. In: U. Pasero & A. Gottburgsen (Eds.). Wie natürlich ist das Geschlecht? Wiesbaden: Westdeutscher Verlag: 199-212

Uncertainty as a barrier in job decision making process of young women

Kathrin Gräßle

Abstract

There are less women working in SET (science, engineering and technology) than men. In my study I show the barriers preventing them from choosing SET subjects when choosing their courses (Gräßle 2009).
There are two main barriers that interact and reinforce each other. The first is the emotion „uncertainty". When choosing a degree course female pupils want to feel secure as soon as possible, as my study shows. The research paper shows, that there are different strategies of decision making. They all don't lead to a technical option. As the consideration of a technical course is regarded as abnormal, the young women dismiss it. This corresponds with the second reason: The incompatible images between being a woman, and their image of technology.
In this article I concentrate on explaining the impact of uncertainty, because it is a new finding that is not described in other relevant studies. By knowing the effects of uncertainty on female teenagers you could find ways for them to enter engineering degree courses they often ignore.
Long-term professional mentoring of processes of choice of study combined with continuing encouragement and reinforcement of technical interests of young women support these processes together with a stimulating reflexion about external influences and self-assessment. This way of coaching is appropriate for schools because for many years educated pedagogues have worked closely with pupils. Institutions for higher education and agencies for employment offer complementary measures for choice of study.

1 The issue: Which are the barriers that prevent young mathematical-interested women from studying technical courses?

The question is, which are the barriers that prevent young mathematic-interested women to study engineer courses?
We know there are less women in SET-fields than men. This has been the fact since SET started as academic disciplines. During the last about 20 years there have been more and more efforts to change this fact. Recently the efforts are becoming stronger, because of the lack of engineers. Currently it is

not only a reason of gender fairness but an economic interest to get more women engineers.

From my point of view arguing economically to get more women in engineering is problematic. Considering economic reasons only neither meets the expectation of women's social participation nor women's integration in the engineering profession. Besides, it must be assumed that the motive will have effects on didactic conception and implementation of measures for attracting women for SET.

Many institutions make arrangements to get women SET-interested. I focus on the activities of universities, and concentrate my study on technical science/engineering. My first approach was to find out why their activities are so ineffective. But before I was able to answer this question, I had to know more of the study decision making process of young women. In my study I want to show how the decision making process of young mathematical-interested women takes place.

First, knowledge from other studies will be presented. General information stems from representative studies from HIS (Hochschulinformationssystem) about choice of study and training and general information about choice of study of a cohort. Circa 2005 the age cohort of investigated pupils HIS stated:

- Over half of the pupils don't feel informed.
- Professional advising and information services are seen critical.
- Information from personal environment is seen as less supporting.
- Services from higher education are strongly used.

There exists, then, a high information and advising need (Heine/Willich 2006).

Even though the study differentiates less between information and decision it shows how difficult choice for study and training are.

Two studies from Hildegard Küllchen and from Kristin Gisbert concentrate on choice of study of young women. Küllchen (1997) investigates the status passage from school to training study respectively.

She states that girls have problems of orientation and decision about profession. She explains this result in terms of the lack of role models that could be identified with that combined professional and private life for women with SET interests (Küllchen 1997: 328). These difficulties are enhanced by gender untypical interests. Küllchen describes three possible reactions:

- Choice of a gender untypical study with the consequence of living in a minority position even in profession
- Choice of a gender untypical study but with gender typical professional career (for example becoming a teacher)
- Time lag of choice of study (Küllchen 1997: 328f).

Difficulties in choice for study are resulting in feelings of hardship and fear of the future (Küllchen 1997: 328).

The work of Kristin Gisbert (2001) shows a connection between choice of study and identity development, especially gender identity too. She concludes that to integrate the feeling of uncertainty going along with a gender untypical choice of study a completely developed identity is necessary. From a former study she knows that for girls in adolescence and women, identity and subject orientation are central for choice of study (Gisbert 1995: 199).

Anne Schlüter (1992) found that daughters of working-class parents choose their study by principle of exclusion and to distance themselves from so-called women' subjects/humanities, but they not aware of these principles of decision. They would say that they make thesedecisions spontaneously (Schlüter 1992: 108). They are not conscious of their choice.

If I can show their considerations and thoughts during their last school years, it will be possible to reconstruct which circumstances, persons, attitudes, activities and information are important to them.

In this way I want to find the answer to my lead question: Which are the barriers that prevent young mathematical-interested women from studying technical courses?

2 The methodology: Qualitative, longitudinal study

The first step of the research was a statistical overview of the German situation of young women intending to go to the university and of the SET-disciplines at the universities. With this background knowledge the sample for guided qualitative interviews was made. The method follows Andreas Witzel and others (Witzel 1985).

Ten female pupils from the 13th grade at German Gymnasium or Gesamtschule were chosen. All of them were interested in mathematics or physics and all had decided to go to the university. The study was longitudinally constructed. The interviewees were spoken to for the first time at the beginning of their last school year and on the second time after their final school exam (*Abitur*), after having decided their further studies and career.

The interviews were analysed using the grounded theory method (Strauss 1998). Patterns and strategies have been taken as hypotheses from interviews. In a second step they were reflected in all cases, verified or falsified, sharpened and structured. To validate the interpretations they were discussed in a science colloquium and in team work with a scientist colleague. The logical argumentation is also important, as it leads to the results and answers the question, why young women don't focus on engineer courses?

3 The result: Strategies for overcoming uncertainty

All interviewees mention experiencing the emotion of uncertainty during their decision making process. Many of them can't describe this feeling while they are looking for a field of study. However, after they took the decision – in retrospect – they talk about their uncertainty. In addition, they were found to practice strategies to forget this uncertainty very quickly. So the strategies I observed are not strategies for study choices, but those to overcome uncertainties. Bearing this in mind, different perspectives on necessary support measures had to be developed. But first the strategies taken will be reflected on.

The uncertainty is founded/motivated in the awareness of the importance of the decision. The young women feel that their choice is going to determine their future lives. And while thinking of this, the determination appears to get stronger and stronger and therefore more and more important. Another aspect of uncertainty is the impression that the decision is irreversible. This leads to feelings that it has to fit, it has to be right, which in turn increases the pressure. And finally, they also feel to be under time pressure. When the school will be over, they have to know what they want to do.

This kind of uncertainty motivates the female pupils, who want to make their decisions in a good and serious way. At a first glance these strategies seem to be decision making strategies. But in my thesis it is shown – across the two interview results – that the stimulus for acting is the uncertainty. The young women want to overcome this feeling.

This different perspective on the situation is very important for researchers, such as myself, who want to find out how we could influence the decision making process in a way to show them SET-disciplines as an opportunity.

In the following I am going to describe different strategies: Their kind, their negative impact to the choice of SET-disciplines, and a method to overcome it. There are three kinds of strategies: Those concerning the method, the social aspects and the habitus. We start with the method-strategies.

3.1 Methodical strategies

Some of the decision makers structure the information they get within a system. They structure by content. Or they structure by assessment, whether it is right or wrong. They hope to find out the right discipline not regarding the possibility of development. Or they structure chronologically. "To structure" is not a bad or a good strategy. To open it for SET-opportunities the system and criteria behind the structuring have to be reflected. Otherwise they work as barriers. By reflection barriers could be overcome.

"To fix" an idea gives security. But without explanation it closes for all alternatives, e.g. SET. To avoid a too early fixation, young women need guidance.

Some of the young women focus their decisions on what they don't want to do. They exclude some of the possibilities. Because they are influenced by image and gender aspects of SET they exclude SET-disciplines. To overcome this barrier they have to reflect on their criteria and their image of SET.

Another similar strategy is *to blind aspects out*. But by blinding it out the young women don't even think about they are not considering it as an option. Instead, they simply ignore it. If they blind SET-courses out, it is very difficult to convince them to open up to new perspectives on SET-opportunities.

Procastination in this case means to put a year on hold. It doesn't mean that they aren't engaged in the decision making process. With this extra time they get the chance to look around and to experience some new fields.

Technological colleges (The *Technikum)* would be a possibility to try something out. There are lot of arrangements for pupils to try universities out. The young women use this strategy to feel active, and by doing so they get the impression that they are doing something. They calm themselves by trying it out. Unfortunately this is not the basis for opening minds. The activities have to be a part of decision-making-system with activities, reflection and guidance.

"To reflect" was mentioned several times as a method of paving the way for SET-disciplines. But reflection is a difficult matter if there is to be a constructive result. The danger is that it can just involve chewing over the problems. Therefore pupils have to learn constructive reflection towards getting a concrete result and reaching a conclusion. Again they need guidance and stimulation for asking questions. Questions like: What are the reasons? What do I feel? How do I wish to live? And they need interaction with some kind of coach.

3.2   Social strategies

A second category of strategies to avert uncertainty are social solutions.

There is already the strategy to communicate with friends, teachers and parents as the first of the social-strategies. The support of those people gives the decision makers certainty. The problem for SET-opportunities is that it depends on the attitude and experience of the dialog partner. If they haven't any contact with SET-disciplines, they can't give such an input. Therefore we shall look at SET-scientists and teachers, who meet these young women.

Elder sisters or friends (especially female friends) are used as role models. Their activities are imitated by the female pupils. The thought behind this is: "What they did was right, therefore, it can't be wrong for me." Again, as I explained in conjunction with "to communicate", the effort for SET depends on the SET-affinity of the role models. It seems to be very difficult to build intentionally SET-role-models, who are personally accepted. This problem can only be solved by a growing number of women in SET.

Young women seek out advice from their parents, their teachers and consultants. They feel affirmed, when they follow this advice. Advice can be helpful, but this is not assured. Again – like in all other social strategies – the environment is important, whether the SET-aspect wins or not.

3.3 Habitus-affected strategies

A third kind of strategies are the habitus-affected ones.

"To be self-confident" is in fact a strategy that brings certainty and could lead to a SET-choice. This indicates out that the young women need affirmation to strengthen them in their present identity, which probably isn't "mainstream". Professionals as teachers and consultants should consider this.

Not only hard facts but also emotions lead the young girls, this is the strategy of being guided by emotions. When emotions become important the emotional image of SET counts. Therefore we should show the practical use of SET. Not only SET for SETs sake, but for the sake of human beings – to put it solemnly.

If the young decision makers just do what they are told to do – that is, that they are solely rule oriented, then there is no basis for new and individual ideas. They need a push to reflect and to try out different things.

Some of the young women argued pragmatically. They mentioned reasons such as that the course should be nearby their parents' home. Just doing what is pragmatic is in some ways a good approach to SET-disciplines, because they are pragmatic disciplines, there are a lot of jobs and good opportunities to earn money. However, many antagonistic strategies mean that being pragmatic is not enough.

The habitus of being fatalistic leads to inactivity and does not create new opportunities. These young women need encouragement and a desire to shape their own careers. That could be achieved by a good guidance.

4   Conclusion: Focus on emotions!

Following these research results, attention must focus more on feelings of insecurity and crisis as a norm. This is necessary in order to attract young women for SET as most of the young women lack strategies to overcome uncertainty as a barrier to choosing SET.

First, uncertainty must be recognized and second it must be handled didactically. A basic condition for approaching the problem of lacking specialists in SET is that technology, physics, and informatics must be subjects in the lives of young people.

Learning to handle uncertainty is part of decision learning. These are not everyday decisions, like choosing a winter coat, instead these are complex decisions for life. Choices around training and where to live and with whom are connected to these decisions.

This work with young people is a permanent and intensive process, which I see teachers as being responsible for.

Schools are also responsible for organization the subjects of technology, physics, and informatics.

Organizations of higher education, agencies for employment and other institutions can contribute toward attracting young women for SET. Higher education organizations are already successful with summer schools, internships and laboratories for pupils. But they should also support young women on an emotional and personal level.

The new approach, that my study showed, is that we have to consider the emotional aspect of the decision making progress. It is important to take all the emotions seriously, both uncertainty and more positive emotions. The positive emotions in this situation are the joy of experiencing something new and the feeling of being the creator of one's life situation.

References

Gisbert, Kristin. 1995. Frauenuntypische Bildungsbiographien: Diplom-Mathematikerinnen; in: Europäische Hochschulschriften, Reihe 6: Psychologie, Band 501. Peter Lang: Frankfurt am Main, Berlin, Bern, New York, Paris, Wien

Gisbert, Kristin. 2001. Geschlecht und Studienwahl. Biographische Analysen geschlechtstypischer und -untypischer Bildungswege. Waxmann: Münster, New York, München, Berlin

Gräßle, Kathrin. 2009: Frau Dr. Ing. Wege ebnen für Frauen in technische Studiengänge. Barbara Budrich Verlag: Opladen, Famington Hill

Heine, Christoph & Willich, Julia. 2006. Informationsverhalten und Entscheidungsfindung bei der Studien- und Ausbildungswahl Studienberechtigte 2005 ein halbes Jahr vor dem Erwerb der Hochschulreife. HIS: Forum Hochschule 3/2006: Hannover

Küllchen, Hildegard. 1997. Zwischen Bildungserfolg und Karriereskepsis. Zur Berufsfindung junger Frauen mit mathematisch-naturwissenschaftlichen Interessen. Kleine Verlag: Bielefeld

Schlüter, Anne. 1992. Arbeitertöchter des Ruhrgebiets im Studium. Naturwissenschafts- und Technikkompetenz und sozialer Aufstieg – oder „Obwohl Papa Schlosser war, haben wir Kinder studiert!" Eine Exploration; in: Schlüter, Anne (Hrsg.): Arbeitertöchter und ihr sozialer Aufstieg. Zum Verhältnis von Klasse, Geschlecht und sozialer Mobilität. Deutscher Studien Verlag: Weinheim: 82- 23

Strauss, Anselm L. 1998. Grundlagen qualitativer Sozialforschung. Fink: München

Witzel, Andreas. 1985. Das problemzentrierte Interview, in Qualitative Forschung in der Psychologie, edited by Gerd Jüttemann. Weinheim: Beltz Verlag: 227-255

The gender perspective of young people's images of science, engineering and technology (SET) in the Slovak Republic

Nataša Urbančíková and Gabriela Koľveková

Abstract

The paper briefly describes the state of the art within the Slovak education system. A special feature of this paper is that it points out the scientific footprints in the rebirth of democracy in the new state. The results of the Motivation project were presented in the paper, particularly the so-called habit of "gift" and the habit of "credit". In the conclusion the reader can find recommendations based primarily on project results.

Introduction towards the gender perspective of young people's images of SET in the Slovak Republic

The motivation for writing this paper was the Motivation project[17] and presenting its results. The following text gives an overview of images of SET found in the Slovak Republic in the research in 2009. The results were found in the area of media influence. The younger generation is surrounded by various magazines and TV shows, which is to a great extent a big influence in their life and future job decision making. On the other hand, the peer group influence is evident and teachers' influence is declining significantly. Family and others (e.g. job advisors) do have an impact on pupils' image of Science, Engineering and Technology (SET). The images of SET in the lives of young people are a complex issue, which is important as it is the new generation that will create the future living conditions.

---

17 Motivation – „Promoting positive images of SET in young people under gender perspective" is funded as Coordination Action within the 7th Framework Programme of the European Comission (Grant Agreement No. 217843) The following people were involved: Felizitas Sagebiel (coordination) and Jennifer Dahmen in Germany, Anita Thaler and Christine Wächter in Austria, Anne-Sophie Godfroy-Genin and Cloé Pinault in France, Els Rommes, Karen Mogendorff, Mariska Schönberger in the Netherlands, Carme Alemany in Spain, Bulle Davidsson in Sweden and Natasa Urbančíková and Gabriela Koľveková in Slovakia and also Minna Salminen-Karlson. For further information see www.motivation-project.com.

Objective and method

The objectives were as follows:

- To look for the presentation of SET & gender in two magazines titled: "Mladý vedec", "Kamarát (Friend)". It was differently presented in the text, but especially in the pictures.
- To look for patterns and attitude in presentation. To find a common style "rule", in use of colours and different motivation "habits" (incentives) in the approach towards the youth audience of the two magazines
- To observe the presence of SET image & gender in the soap opera "Panelák (Block of Flats)". It is performed light humour in line with Slovakia's history of soap opera production. (There was no expectation of any special focus on SET at the beginning).
- To examine two specific secondary schools (one "girls" and one "boys" school), i.e. two case studies comparing curriculum and attitude towards mathematics.
- To examine two case studies, to compare images of SET, and the relevant adjectives from the pupils' interviews.

As for the methods: The first was the literature review up-date. Secondly, the magazines were selected for research, i.e. the most popular magazines and TV shows/series in the age group 14-19. Afterwards the picture analysis followed in line with guidelines (Thaler 2009a). Thirdly, the teacher's interviews were arranged. Teachers answered 23 questions, although the interviews were semi-structured. Fourthly, interviews with pupils were arranged and carried out.

Outcomes

The Slovak education system

From the history of the development of education in the Slovak Republic, the following milestones should be mentioned:

- Constantinus and Methodius in 9th century for their role in education in the Old Slav language.
- Matthias Corvinus founded the first university called Academia Istropolitana in 1465.
- Catholic and Evangelical schools were founded in the mid.16th century (e.g. Premonstrates).

- Ostrihom Archbishop Peter Pázmany gave impetus for the foundation of Jesuit University in Trnava, with many faculties (some faculties in Košice) in the period 1665-1769.
- Thanks to its technical orientation and scientific research the Mining Academy in Banská Bystrica (founded in 1763) became known worldwide.
- Maria Theresa founded Collegium economicum in 1763.
- The modern education system was established via Ratio education in 1777. In 1805 the new "Ratio" required school attendance from 6-12 years.
- The so-called Thuna Education Reform (suggested by F. S. Exner and H. Bonitz) in 1850 brought the development of general schools (6 years of secondary grammar schools changed into 8 years).
- With the raise of the Czechoslovak Republic (1918) the educational policy unified the system.
- Comenius University in Bratislava 1919 arose. In 1937 the Technical University of Dr. M.J. Štefánik in Košice and in 1940 The Commercial University was established in Bratislava.
- After 1948 Czechoslovakia started to build up the education system based on the Soviet models (state-owned, unified schools, the education system has involved on-the-job training).
- Transformation of education system began after the Velvet revolution in 1989. Decentralization is the main feature of the reforms that took place including the up-dating of the content of education.

Although the transformation has been ongoing for 20 years, one can find the headstones of the education systems that had been preserved as a good base over the years (e.g. 8 years gymnasium as an opportunity to study). Nevertheless, the 40 years of socialistic development cannot be erased; therefore one should be thankful for what has been built over the years. On the other hand, one need to proceed further from the standpoint one has. The basic elements of transformation were the "educational programmes". According to these programmes the education and training has being carried out. There exists the compulsory curriculum (through the State educational programme). The Curriculum Board is a special initiative and advisory body of the minister of education. The National Institute of Certified Measurements of Education has its role in monitoring improvement of education process. Further agents of the youth education are: Slovak Committee of Parents's Associations, Pupils' school board, School board, municipal school board, territorial school board, school self-governing bodies, National Institute for Education (ŠPÚ), National Vocational Education Institute (ŠIOV) Municipal corporations and local bodies of state administration

The following research results are applicable to the ISCED 3 level of educational system

Scientific footprints found in the print media

Socialistic footprints were found in the media, which had been facing the change in the rebirth of democracy in the "new state". The change was gradual, not abrupt. Although, looking back over the 20 years (since 1989) one could admit it was a quick turnover compared to other transition processes in the community.

We were analyzing the unofficial sources of job information, i.e. the emphasis was given on the following magazines published in Slovak language:

- Mladý vedec
- Kamarát

Table 1: Description of the magazine Kamarát and Mladý vedec in respect with criterias

	Magazine "Kamarát"		Magazine: "Mladý vedec"	
Period	2007	2008	2007/08	2008/09
Circulation	39 018*	43 781*	15 000**	35 000, 25 000**
Audience	13 - 19 years old		6 years and older**	
Scientific footprint	The footprints are more subtle, seen in advertisements, competitions are focused on winning a product as a gift. "A gift habit." – a passive approach***		The footprint is obvious, the competitions are focused on winning a "prestige". "A credit habit." – an active approach ***	

Sources: * *TREND Holding 2004-2008. ISSN 1337-0006. Tlač. Grafy a tabuľky. On line [9.12.2008] Available at:*
http://medialne.etrend.sk/tlac/grafy-a-tabulky.php?medium=tlac&co=mesacniky
** *Correspondence with Ing. Mgr. Martin Hriňák – coordinator of the project Mladý vedec,*
*** *Own elaboration, explained in the text*

The magazines were chosen for the analysis due to several facts. The first criteria of selection were especially popular among young people. This was supported by the statistics on circulation or subscriptions to these magazines. The second criteria was the target audience of the magazine, which was young people between the ages of, on average, 13-19 years old, and the gender difference (if there is one in the magazines). Magazines for young people usually do not distinguish between girls and boys; they present themselves as gender neutral, because their objective is higher profit by having more readers (no matter girl or boy). A third criterion was the scientific "footprint", if there is a special attempt to work with young scientists via media.

Some details about the magazines are summarized in the table 1.

Obviously, magazines have been forced to struggle within the market powers of competition over the years. Magazine *Mladý vedec* got financial support from European Social Fund at the beginning. One could predict that their marketing strategy was going to change in order to obtain funds from advertisements, which are usually conditioned by the number of readers approached. One can assume that the *Mladý vedec* can muddle through in this way, because there are plenty of companies and devices to advertise. Furthermore, the employers looking for good future employees can start seeking them with help from the competitions in these magazines also using problem tasks for solving their problems.

The magazine "Kamarát" has struggled its way through the transformation period from the planned economy to the market or mixed economy. Thus, the managers of the magazine know how to react to the needs of the consumers.

As for the pictures and content analysis – the pictures are tributary to:

- The advertisement desired or to the message they have to get out e.g. science,
- the strategy, mission, goals and focus of the magazine editors. In Slovakia it is rather gender neutral or balanced.

Picture 1: The picture shows cooking is a "female – woman's job".

Sporák so sklokeramickou varnou plochou

Source: Slovakia_11, Magazine Mladý vedec, number 3, April 2008, ISSN 1337-5873, page 25

The feature of the pictures is competition, but in the sense of "endowing" gifts or in the sense of deserving or achieving "prestige". So there were two approaches uncovered:

- "Passive" – just take the gift, if you are lucky or
- "active" – give and take, prove and give your knowledge in the competition and take your fame or medal.

Further, the pictures of winning teams are also supportive of collaborative spirit. Some stereotyping in the pictures persists. The first picture (see picture 1), coded below as Slovakia_11, shows women in a traditional role, cooking. But this is stereotyping to an extent only,as there are numerous pictures, such as the one below coded Slovakia_12 (see picture 2), that show women working in the woods with the text below explaing that she is carrying out geological work.

This woman is using the cooker and she is touching the hob with her right hand, which shows the fact, described next to the picture, that the hob is not hot as the cooker uses "turboefect" etc. Cooking is a "female – woman's job".

Picture 2: The picture shows a women working in the woods.

Meranie kompasom patrí k základným zručnostiam geológa

Source: Slovakia_12, Magazine Mladý vedec, number 3, April 2008, ISSN 1337-5873, page 28

According to the text under the picture the women on the photo is using a compass for measurement, which is claimed "to belong to the basic skill of the geologist". This picture is not stereotyping.

Magazine Mladý vedec

It started to appear in 2007. Thus it is relatively unknown. It started to be distributed (see table 2) for free as a result of a project financed by the European Social Fund. The aim of the editors was to provide the news from the technical and natural sciences. Plus a special focus was given to competitions.

Table 2: Concerning magazine „Mladý vedec" – subscription and circulation statistics for 2007-2009

Period	2007/08	2008/09
Number of subscribers	13 593	24 379
Circulation	15 000	35 000, 25 000

Source: Correspondence with Ing. Mgr. Martin Hriňák – coordinator of the project "Mladý vedec"

The readers of the magazines are people age from 6 to 70 (even above), while the principal group are the readers of age from 10 to 19, as stated in the correspondence with Ing. Mgr. Martin Hriňák, who is a coordinator of the project "Mladý vedec".

What is a striking fact is that this magazine is distributed among:

- Elementary schools (700)
- Secondary grammar non technical schools (100)
- Secondary schools that are together technical and non-technical ones (133)

The magazine creators have cooperated with Ministry of Interior of the Slovak Republic and publishing house Perfekt and they have older issues of magazines "Quark" and "Fifik" to distribute free of charge with the new magazine "Mladý vedec".

The "*Mladý vedec*" was distributed for free and teachers in particular got to use it as a supportduring the lessons or for organizing the competitions. The magazine *Mladý vedec* itself organizes a correspondence competition for its readers. In order to prove the readers' potential the statistics of competitions participants can be added. Concretely, 448 were girls out of total number of participants 854. That means that more than half of the participants were females. The magazine existed for ashort time only, thus it was not possible to see a difference in female participation across the years of publication. Before 1989 there was similar magazine called "Veda a technika mládeže" – translated as "Science and technology of youth", which was aimed at the students aged 15-19. There was also a magazine "ABC" aimed at pupils aged 6-19. This magazine was published in the Czech Republic and is still published and used to be extremely popular.

For the magazine *Mladý vedec* we can strictly divide the pictures into three categories:

- Pictures of scientific issues (e.g. An analysed picture from magazine *Mladý vedec*, number 5, April 2008, ISSN 1337-5873, page 2. The picture was showing the "intelligent plasticine", its preparation etc.).
- Pictures with persons and technique (e.g. An analysed picture from magazine *Mladý vedec*, number 3, April 2008, ISSN 1337-5873, page 21. This picture is showing a microscope that can be hidden in the pocket. The picture was in the magazine section "curiosities from the science and technology". It shows women using it in the background in a chemistry setting, which is considered appropriate in terms of women's attitudes to science, i.e. that chemistry is considered appropriate for women. Picture provided below.).
- Pictures of winners or teams. (e.g. in those pictures girls are standing in the front line).

Picture 3: The picture shows "intelligent plasticine" and its preparation.

Source: Slovakia_7, Mladý vedec, number 5, April 2008, ISSN 1337-5873, page 2.

Picture 4: This picture is showing a microscope that can be hidden in one's pocket. The picture was in the magazine section "curiosities from the science and technology". Women are shown using microscope in chemistry background, which is considered to be an appropriate science for women.

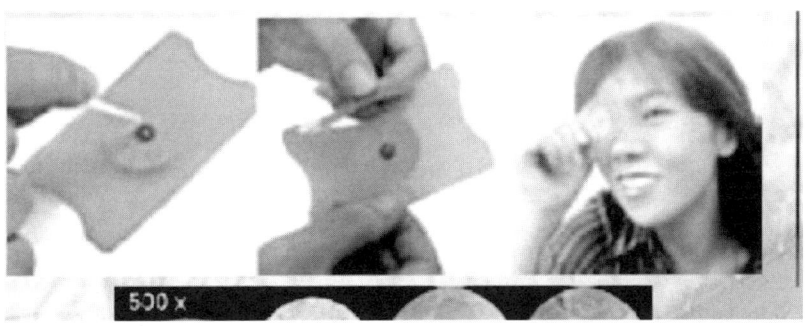

Source: Slovakia_10, Magazine Mladý vedec, number 3, April 2008, ISSN 1337-5873, page 21

The rest of the pictures of people were pictures of winners or teams (see pictures 5 and 6).

Picture 5: Winners

Source: Magazine *Mladý vedec*, number 3, April 2008, ISSN 1337-5873, pages 29, 35-36

As this magazine is really science oriented most of the pictures in the issue are concentrated on facts. The authors were focused on explaining factual subjects and not on "increasing the number of issues sold." Nevertheless, the magazine is colourful and nicely designed. IMPORTANT: As they have a different focus, they must have a different strategy and the magazine looks different to the second analysed magazine *Kamarát*.

Picture 6: Winners of a Physics Olympiad. Girls are in the front line.

Spoločná fotografia ocenených žiakov

Source: Magazine *Mladý vedec*, number 3, April 2008, ISSN 1337-5873, pages 29, 35-36

Magzine Kamarát

Kamarát magazine was selected because it is similar to *BRAVO* (the largest teen magazine within the German-language sphere). It is more "Slovak-like" and it is preferred to the foreign publication BRAVO.

Table 3: Concerning magazine „Kamarát" – selling statistics for 2006-2008

Period (January)	2006	2007	2008
Sold (pcs)	44 475	39 018	43 781

Source: TREND Holding 2004-2008. ISSN 1337-0006. Tlač. Grafy a tabuľky. On line [9.12.2008] Dostupné z: http://medialne.etrend.sk/tlac/grafy-a-tabulky.php?medium=tlac&co=mesacniky

The *Kamarát* is one of the best selling monthly magazines, see table 3. In October 2008 there were 41,041 issues sold per month (according to www.medialne.sk). It is thus the sixth best selling monthly magazine since 2006. The results were similar to ones available from the Audit Bureau of Circulation. This office does not yet monitor *Mladý vedec*.

Comparing the circulation numbers of *Mladý vedec* and *Kamarát*, it was clear that *Mladý vedec* had, after only approximately two years, acquired half of the circulation of the *Kamarát*. The fact that it was distributed free of charge helped to establish communication channels with readers. *Kamarát*, however was being published since before 1989 and has survived until today thanks to its ability to adapt in terms of "image", style and topics. Such an adaptation was noticed also in case of the above mentioned magazine *ABC*. But it was not the case of the magazine "VTM – Veda a technika mládeže", which had disappeared. Plus the editor and all writers and other people were responsible for preserving the existing magazines. In the market economy the magazine provides what is demanded, thus supply is in response to customer wishes. In the planned economy there were laws regarding supplying the customer with what (s)he wanted in terms of what society needed or "agreed" on as being most suitable. Thus, adjusting the demand and shaping the wishes of young people as well as setting "some rules" for young people.

Summing up the images of SET identified in these magazines, we can conclude that: Science is fun, competitive, one has to have a winning ambition, and science is a studious subject, and an issue in advertising.

Scientific footprints found in the TV soap opera

The "Panelák" soap opera, which is broadcasted on a daily basis (Monday to Friday), was selected for the analysis of its images of SET. The meaning of the title can be translated as "Block of flats". In line with the guidelines by Thaler, A. – Scheer, L. (2009b) five episodes had been selected between March and May 2009. The titles of the episodes were:

- "The Fiance" (March),
- "The Grandmother" (April),
- "The Commission" (April),
- "The Disclosure" (May),
- "The Trial period" (May).

Each episode was approximately 35 minutes long.

The general theme of this soap opera could be derived from the credo "Miesto, kde to žije." i.e. "The place which really lives." It describes the life of families living in a Block of flats. A spectator can share the happy and sad parts of their life via TV. One can see older, middle-aged and the younger generation taking up the challenge to live in the world of "today" situations.

An important aspect of the webpage is an interactive section, which is designed as a children's room, most probably the twins boys Andrej and Janko. Children can do the following activities:

- Colour in pictures
- Create their own strip cartoon, which you can later broadcast. There have been around five thousand of these published (as of May 2009) in the so-called "book-case"
- Solve a puzzle

Also the soundtrack song of the TV soap opera is quite poplar.

A description of all of the soap opera's characters can be found at the website www.panelak.sk. But we will focus here on one female character, Jarmila, who does represent some "scientific footprint" in her behaviour.

Jarmila Švehlová (69 years old), mother of Jakub is living with the young couple (Jakub and Ivana). Jarmila has never been married. She moved from the Czech Republic to Slovakia, when Jakub was 5 years old, she used to work in the army, thus moving was a normal regular aspect of her lifestyle. As an "army woman", she is strict and vigorous and realistic. Her personal and professional past is a mystery. She has got lots of information about the happening in the world and everywhere, there is nothing that surprises her. She can forecast things that will happen. She likes Ivanato keep the flat orderly as this helps her to handle situations somehow. Jarmila is a medical miracle, who is doing better and better since she was told she has cancer of the stomach (10 years previous) and Alzheimer (5 years previous). This fact is annoying for Ivana. Jarmila has finished university thanks to her enrolment in the army. She studied while she was working in the army. She was working as telegrapher or operator, then in counterintelligence, where she achieved the rank of major. She seems to know a lot, but never says a lot, she only reveals tiny bits of information. She seems to have not gotten married, because in her position in the army it was more convenient. She does not want to reveal who Jakub. Blichár tells Jakub that his father could have been someone important; because the day he was born Svehlová was promoted twice. Jarmila's hobby is to model airplanes and while doing this she listens to the TV. Her knowledge of army technology is at an excellent level. She does not like Blichár at all. Concerning her celebration days, she asks Ivana, Jakub and friends to get her a new airplane models that she does not yet have in her collection, which makes a perfect and "easy to think of" present for her. She must always have some stock of models, she enjoying gluing them together, it takes her about four weeks to glue one model, it is a passion for her.

Concerning the rest of the characters, the context makes some of them a typical "SET actor". Characters that promote an image of SET image do feature. The characters actually explain some motivational factors as to how a person starts his or her interest in SET. For instance:

- Bajzová – is on maternity leave for a long time and she has always been clever since childhood. Thanks to her children – twins – she likes computer games and gets to play with technology she had not known about before,
- Kordiak – contrary to Zuzana Bajzová he does not use new technologies (but he does own them), based on an adolescent passion for music he prefers using older technology – gramophone,
- Švehla is a passionate adolescent, he is keen on IT, he enters competitions that should have improved his self-confidence, but only result in proving his expert knowledge,
- Jarmila Švehlová – thanks to her work in the army it was possible for her to evolve the passion of telecommunication and airplanes as well as other technologies.

One can see Jakub Švehla is taking after his mother Jarmila, he is skilled in technology. He had a role model albeit it a female role model. Švehla is not an easily understandable person because of the different vocabulary he uses e.g. generating streams or catching bags. If one was to rank the number one "SET actor" of the soup opera: Then Švehla and Jarmila Švehlová are clearly at the top of it.

The female images of SET presented by characters were noted as follows:

- Female images:
- Zuzana Bajzová learns from her children about technology – games are mainly realistic games or show images of "games technology" that bring together the generations of different ages,
- Jarmila Švehlová – represents images of "work" in the army as a mediator of technology without gender discrimination and images of a "hobby" that is considered unusual for woman, i.e. making model airplanes.

Should there be any recommendation towards producers and authors concerning the SET image then they could change some of the storylines from time to time. Authors and producers may have some ideas regarding the webpage of Panelak as there are plenty of cartoon scenes prepared by the audience themselves, as discussed above.

The research identified several images of SET from the perspective of characters. Characters reflect the personal image of SET, which is tied in with SET occupations: IT specialist, doctor, retired scientist. SET image could be assessed as strong especially for mobile phones and music equipments. The way in which SET images in media were used included: To communicate to escape the loneliness, to get the job done, to entertain oneself, to improve the live style (coffee, music). Nevertheless, it is not easy to prove what the spectators of the show can absorb and cannot see or comprehend. A special character was Ms. Svehlova (69 years old), who was labelled as an "SET keen" person, although from an older generation. This relation to the "old days" was to recall the socialist era time when it was appropriate for women to use and to study technology or be a scientist.

Scientific footprints found in the interviews with pupils and teachers

Two schools were examined, thus there were two case studies. The first case study school was predominantly for boys, but girls were obliged to attend it between 1945 and1989 (girls made up around 30% of the student population). Later the portion of girls fell (to almost 1% according to internal school data). The second case study was a school, which was predominantly for girls for the years 1945-1989 (boys made up only 10.5% of the student population). Later the proportion of boys increased to 30% (internal school data).

To compare the curriculum between the first case study and the second case study in brief: There are lots of technology subjects in the first case study and there are some technology subjects in the second case study. This is important in terms of school selection, which proves that images of SET are not popular among girls, or that they seem to be scared away from supposedly "difficult" subjects. One recommendation here could be to promote of "light" technology subjects, where anybody can score good points and achieve good marks. The bad marks could prevent in both case studies pupils having difficulties in enrolment at university. Concretely, maths is a subject taught at both case study schools. The number of lessons during the week is the same. The status of maths is objectively the same, though subjective judgments of pupils and teachers were little bit different. In the first case study maths is accepted as a good help for understanding and following the other subjects. While in the second case study maths is considered as a subject that is not needed too much later. It is accepted mainly as important for later university enrolment (if considered). Therefore, some students are taking private lessons in maths.

One of the outcomes from the students' interviews was summarized as the obtained adjectives with SET.

Some adjectives associated with SET from first case study were: Interesting, functional, important, helpful.

Naming some adjectives associated with SET from second case study included: Helpful, interesting.

Comparison between the two case studies' adjectives and answers context shed a light on the difference between SET images and liking SET: The pupils from the second case study did not link the toys (e.g. the play "Man do not be angry") with science or technology, whereas the first case study pupils did (e.g. PC games). The pupils in both case studies associated the SET images with being helping (whether furniture, remote controls etc.).

To conclude, without presenting the rest of an abundant research material, the hypothesis concerning teachers, job advisors relation, and some dreams/nightmare information has led to the following results:

- To promote an image of an "SET brother or sister" (in family or otherwise),
- to promote "real life" activities (esp. by teachers),
- to promote endogenous motivations – initiatives, outside the school (e.g. job advisors),
- to promote the linkage of subjects (teachers),
- to promote a positive attitudes overall – active approach, a habit of credit as expressed above (cooperation of school, magazine editors, families etc.).

Recommendations towards the "motivation agents"

The media analysis was intended to prove the country's post transitional footprints labelled as "a gift habit" and "a credit habit". The peer group influence prevailed over the teachers influence on SET images and job decisions. Moreover, media is a strong influential factor on all young people. The TV soap opera analysis proved that television is reflecting some consequences of gendered public images for SET decisions (e.g. Ms. Švehlová – military worker, Mr. Švehla – her IT specialist son). Teachers were very aware of the dream/nightmare jobs of their pupils.

- In terms of the magazine editors: They used amore "active" – "a credit habit" approach for the children, pupils.
- Regarding producers and authors concerning the SET image: Authors and producers may have some ideas from the webpage of Panelak as there are plenty of cartoon scenes prepared by the audience themselves. The use of the webpage is itself proof of the active use of SET tools, the further development of which we encourage, e.g. in association with magazines like *Mladý vedec*.
- Regarding teachers and job advisors – to continue their diligent work and cooperate with other actors in the field of youth education (editors, TV producers etc.), to promote an image of an "SET brother or sister" (in family or otherwise), to promote "real life" activities during the lessons.

Generally, to promote positive attitudes overall – active approach, a habit of credit as expressed above (cooperation of school, magazine editors, families etc.).

Agreeing with Piscová et. al (2007) all actors ought to contribute towards an increase of "... diligence, studiousness, preciseness and so forth i.e. the features, which are an assumption of work of a researcher ...", which she found to be a missing value among young people.

Conclusion

The SET images were identified as follows:

- In the media: Science in magazines is fun, competitive, one has to have winning ambitions, and science is an issue of studiousness, an issue of advertisements. Characters in TV reflect an SET image, which is tied with the occupations: IT specialist, doctor, retired scientist. SET image could be assessed as strong especially for mobile phones and music equipments. The aims to use SET images in media were: To communicate to escape the loneliness, to get the job done, to entertain oneself, to improve the live style (coffee, music).
- At school: Science is a helpful, functional tool.

The research into SET images was intended to find the motivation for young people, in particular girls, to join the world of science and technology. Based on the images of SET found, one can recommend the other images that could be joined via media, school, family, with existing ones in order to improve the situation. The suggested new images were:

- The image of an "SET brother or sister".
- The image of "real life" activities.
- The image of endogenous motivation.

References

Piscová, M.; Bahna, M.; Zeman, M. 2007. Podoby ženy. Sociologický ústav SAV, Bratislava, ISBN 978-80-85544-52-7

Thaler, A. 2009a. WP2 quantitative media analysis: SET and gender in youth magazines. Motivation-project. Members Area. Administrative part. Guidelines. [16.10.2009]. Available at http://www.motivation-project.com/files/guidelines/WP2_quantitativemagazine_analysis_gender_09_12_2008.pdf

Thaler, A.; Scheer, L. 2009b. WP2 media analysis: Soap analysis. Motivation-project. Members Area. Administrative part. Guidelines. [16.4.2009]. Available at http://www.motivation-project.com/files/guidelines/WP2_SoapAnalysis_AT_16_04_2009.pdf

Secondary school students' reactions to descriptions of engineering and nursing in university catalogues

Minna Salminen-Karlsson

Abstract

The underlying assumption in this study is that it is not only the technical content of engineering education, but also the educational framework that deters young people, and particularly girls from engineering studies. Descriptions of engineering programs and nursing programs in Swedish university catalogues were compared. The results show that both descriptions draw on gendered professional and educational discourses, creating distinctively different descriptions. A questionnaire study among secondary school students showed that when the subject content was removed, engineering and nursing programs were equally attractive to both boys and girls. However, nursing programs in general were more attractive than engineering programs.

Background

This study is ultimately based on the notion that it may not be (only) technology that puts off girls from studying engineering, but that the notion of engineering and engineering education is associated with a number of other characteristics that may seem less desirable to them.

To find out what kind of characteristics engineering programs associate themselves with, this study analyses what is offered to presumptive students in the recruitment catalogues[18] about engineering programs from different universities. In this examination it is not the content of the educational program that is interesting, but the framework – teaching methods, study environment, connection to working life – in which the content appears and what kind of language is used to describe it. The second part of the study examines how secondary school students evaluate such descriptions of educational programs – how girls and boys comment on descriptions of the educational framework and the language, if they cannot relate it to the gendered area of the subject matter.

For this study, not the high status graduate engineering education, but the three year program leading to the bachelor of engineering degree was chosen.

---

18 When this text refers to brochures or catalogues, it is because the actual study was done on university program catalogues. In most cases the same information was available on the Internet.

This engineering program is a relatively new phenomenon, created in the 1990's, and not very well researched. Most research has been done on the graduate engineering programs, which have a much longer history. There are also a number of gender analyses of graduate engineering education (Dryburgh 1999; Sagebiel & Dahmen 2006; Salminen-Karlsson 2003).

The Swedish three year education was a development of a two-year education that, in turn, partly substituted an earlier four year technical program in secondary education. The main difference in the national degree requirements for bachelor engineers and graduate engineers, respectively, is seen to be the phrase 'design and manage products, processes and systems...' for the bachelor level and 'develop and design products, processes and systems...' for the graduate level (The Higher Education Ordinance, Annex 2), even withstanding other differences in the descriptions. In short, the bachelor of engineering education is seen as supplying the industry with engineers who are both hands-on and well educated enough to handle modern, complicated technology. While the two engineering degrees were strictly separated during the 1990's, it is now often possible to continue from a bachelor degree to a graduate degree. This indicates that the hands-on aspect has become less prominent.

The student population of bachelor engineering programs is quite different from that of graduate engineering programs: The students come mainly from the lower middle class (Högskoleverkets årsrapport 2008), they are older (22% are 25 years or older when they start on the program, Statistics Sweden 2009b) and thus, many have some working background before they start their studies.

The two engineering programs can, to some extent, be expected to mirror Wajcman's (1992) description of two kinds of technical masculinities – middle class intellectual technical masculinity and working class practical technical masculinity. This also means that the studies on gender in graduate engineering education do not directly apply to bachelor of engineering education.

To highlight the possible special characteristics of engineering education, a material for comparison was needed. The education for hospital nurses was chosen as the program for comparison for several reasons: The programs are both three-year programs and are offered by a number of regional university colleges. They recruit the bulk of their students from the same socio-economic group. And, of course, both these programs are gender segregated, the percentage of women in engineering programs being 23% and the percentage of men in nursing programs being 15% in 2006 (Statistics Sweden 2009a).

In Sweden, a number of projects have been conducted to make both engineering education and nursing education more gender balanced. However, the ordinary gendered discourses of the education and the professional fields counteract these efforts. The different histories have formed the regulatory and material practices of the different programs. The programs are organized differently, and this naturally shows in program descriptions. But also similar phenomena are described in different ways in the different program descriptions, using different vocabularies, because the different educational programs connect to different discourses (see, for example Perelman 1999 about engineering discourse and Dufva 2004, about nursing discourses in Sweden during different periods). An example of the different vocabularies are words like 'responsibility', 'leadership' and 'stimulating', which are common in presentations of nursing programs but do not appear in engineering programs, and 'advanced', 'competence' and 'exciting', used by engineering education but not nursing education. This is in spite of the fact that nurses use advanced technology and have special competencies and that nursing can be exciting as well as stimulating. Correspondingly, some engineers will have leadership positions, they are also responsible for different things in their profession and engineering is stimulating as well as exciting. Thus, the different descriptions could to some degree be interchangeable, but are restricted by the discourses from which they draw their vocabulary.

Gendered descriptions in the university catalogues

For this study, all university catalogues for the autumn term of the year 2006 from universities offering a nursing program or a bachelor of engineering program in engineering technology or computer engineering were examined. The two engineering programs considered here were chosen because the program in engineering technology is a program in a traditional area of engineering and because the program in computer engineering is even more male dominated than most other engineering programs. The descriptions of these programs were coded with the program Atlas.ti and the following opposite categories emerged as a result of an inductive examination of those codes and categories.

- Engineers are supposed to be interested in the subject area of the education, i.e. in some more or less defined technical area. Nurses are not expected to be interested in the subject area of medicine or even in the subject area of care. They are expected to be interested in people. The subject area of care is introduced in the texts, as the subject area for those who are interested in people. However, in the opening lines of the texts, an interest in people and sometimes (but not very often) helping people is what you are expected to have, to be an addressee of the text and read further.

Studies on gender in engineering have discussed the people-things divide that seems to be so important in the making of engineering identities (Faulkner 2001, 2007; Lagesen 2007). This study shows how this divide is established already in the program catalogues, when describing the area of study for presumptive students. Rommes et al. (2007), find that girls can choose technology as long as they can keep their identity as people-oriented and caring. However, the descriptions of engineering in the program catalogues do not address these individuals.

- Engineers need competencies, nurses need qualities. This is well in accordance with the national degree requirements, where nurses but not engineers are supposed to have personal qualities such as 'demonstrated self-awareness and the capacity for empathy' (Higher Education Ordinance, Annex 2). This means that those young people who choose to become nurses have to be prepared not only to learn knowledge and skills, but also to undergo a socialization process to make them to the right kind of persons – something that is not re-quired of engineers. It is also in accordance with the societal tendency to regard women's skills as innate characteristics and men's skills as acquired competencies.[19] Even the idea of girls being more adaptable and submissive may play in here.
- Engineers work with production, nurses work with reproduction. The division production – reproduction was central for second-wave feminism in their argumentation how both politics and economy have prioritized the productive, public sphere where men dominate, while the reproductive, private sphere that long has been women's has been neglected.[20]

When it comes to descriptions of the educational programs, the link between engineering and production becomes almost over-explicit. This is well in accordance with Perelman's (1999) description of engineering discourse and its concentration on production. According to the catalogues, you can learn about production technology, production flows, production systems, production compilation systems, production strategies, production logistics, production quality, production preparation and high-performance production. Production can be improved and developed. One can specialise in production or industrial production or lean production. During one's education one can work in the production at a company. To be an engineer, one obviously has to have an interest in producing things. The texts do not mention the fact that many engineers actually are, in a way, engaged in reproductive work: Sustaining and servicing things that others have produced.

---

19 See, for example, Kelan 2008, about social skills in engineering.
20 This has been elaborated by, for example, Ehrenreich 1984 and Walby 1990. Mulinari & Sandell 2009, write about how present day societal research still connects women with reproduction and not with production.

The corresponding word for nurses is (nursing) care, which is not producing but sustaining and preferably improving the health of other people.

- Engineers are supposed to deliver quality, nurses are supposed to be responsible. The different sides of the production – reproduction divide can be seen as functioning in different ways and producing requiring different characteristics. One example of this is the idea of an ethic of care, used predominantly by women, in contrast to the ethic of rights, used predominantly by men (Gilligan 1982). Girls are still more often than boys socialized to having responsibility, not only for themselves but also for other people (Wahlström 2004). The female dominated nursing education can be said to continue this socialisation. Responsibility is a word that appears often in the texts about nursing. In texts about engineering it is almost non-existent. Obviously, engineers are not supposed to feel overall professional responsibility. Their approach to their work and what is expected of them is, instead, described in terms of quality. Thus, their 'responsibility' is more particular.
- Still one difference connected to the production – reproduction sphere is the way the need for communicative skills is described. According to the catalogues, nurses need to know how to 'inform', 'instruct', 'tutor', and 'co-operate in groups', while engineers need to learn how to put their message forward to gain support for their ideas (Jang et. Al 2009; Peterson 2008). Interestingly, even though much of engineering texts in the catalogues deal with project work, the importance of learning to co-operate (instead of just co-operating) appears seldom.
- A thoroughgoing difference between the texts is the presentation of the program as an academic study program or as a program close to working life. The texts describing engineering education make it seem as if the academic content were almost superficial, something that has to be there to make it a study. For example, the word 'company' appears far more times than the word 'teacher'. In engineering texts the buzzword is 'project work', which seemingly is the method to learn everything that is needed – even if project work might actually be a minor part of the education, when the text is read more carefully. The texts about nursing write much more about teachers, teaching methods and the subject matter.

In this context it is understandable why the periods of practical work during the education are called 'education in work context' in the nursing texts – it is still education that the student is engaged in. For engineers it is a question of an 'internship', and the texts give the impression that the student will be part of a company and possibly even a useful part of it, instead of being a learner, as the nursing texts suggest.

The engineering texts give the impression that they are targeted at young men, tired of school but willing to make a career. Engineering education is presented as an individual, educational project. In contrast, the nursing texts seem to be targeted to studious girls who do not mind continuing school, willing to both pour over books and to train practical skills under competent supervision.

- Connected to the work – school divide is also the general impression that engineering education fosters individuals, while nursing education fosters members of a collective. The great number of elective courses in many engineering programs contributes to this – sometimes it seems that there are not two engineers with the same education, while nurses are expected to follow a well laid out path through their studies. But the idea of an individual career vs. being a member of a collective is indicated even in other ways. For example, there is a common promise that after an engineering education the student will be 'attractive' in the job market. Nurses, for their part, never become 'attractive' in the texts. Most often they become 'well prepared' or 'in demand'. The engineers are sent out to sell their individual attractiveness, while the newly graduated nurses are a commodity for which there is a demand.

Thus, to be attracted by the descriptions of bachelor of engineering education, it is an advantage if a student is not only interested in technology, even if this is one of the prerequisites. She should also be interested in producing and not servicing, accept that the education does not contribute to personal development but only to development of skills and knowledge, she should accept that responsibility is not a characteristic that is particularly cherished and that communication equals assertion. She should also feel independent and competent enough to navigate through a number of subject choices, to assert herself in project work and to easily integrate in a company environment and after the education go out to a labour market where she will need to sell her individual competence. It is also an advantage if she is not of the studious 'good girl' type, not a reader but a doer.

The recruitment texts for both nursing and engineering seem to be targeted at the (gendered) group now dominating on the program. This is hardly surprising: The texts are advertisements, and advertisements still often take the safe way of using traditional stereotypes (Plakoyannaki 2008; Wolin 2003).

Questionnaire study

Introduction and methodology, questionnaire study

However, young people of today may not conform to old gender stereotypes. There actually may be a fair number of girls, who might feel comfortable with the implicit requirements of a bachelor of engineering program. The descriptions may conform to gendered stereotypes, but does that affect the educational choice of secondary school students? To find out, four descriptions of educational programs were created, none of them stating the subject matter of the program, but only describing the program and the job market in general terms. Two of them used phrases from descriptions of nursing programs, and two used phrases from engineering program descriptions.[21]

These descriptions were distributed to 366 secondary school students, 205 girls and 157 boys[22], on social science and science programs (the two secondary school programs which are the main track to higher education). 117 of the students attended school in a middle sized town, with an industrial history but now also having a university college, and the rest in five rural municipalities. Having a good representation from rural areas was seen as an advantage in this context, as it is these students that many of the nursing and bachelor of engineering programs target.

It was not apparent when recruiting the schools to take part in the study that two of the schools had a special profile in sports, and thus enrolled a somewhat a-typical student body. Some of those girls who were enrolled in the sports profile (the percentage is unknown) may, thus, have other interests than the 'typical' secondary school girl.[23]

The students were asked to read the descriptions, choose one of the programs, motivate their choice and preferably also comment on the other programs (which many of them did). In addition they got a questionnaire with 37 different aspects relevant for an educational program, also derived from descriptions of nursing and engineering education, and were asked to mark on a scale how important these aspects were for their choice of education.

---

21   The only conscious 'manipulation' of the texts was including a phrase about working with people in both descriptions of engineering education. Both phrases came from original descriptions of engineering education programs, but there are still many such descriptions that do not mention working with people, and, thus, a random pick of two engineering descriptions would unlikely have resulted in two descriptions mentioning people.
22   Four students did not specify their gender.
23   For example, becoming a police officer was a popular future option among the girls, and there were also a few who planned to go into the military. Nursing was not a popular choice – however, such typical female professions as teacher and social worker were popular.

Results of the questionnaire study

The results show that even if the descriptions appear gendered, the different programs attracted approximately the same numbers of girls and boys. It was only one of the nursing programs that attracted significantly more girls than boys. However, when motivating their choices, girls and boys drew on somewhat different details in the descriptions.

The two nursing programs attracted significantly more students than the two engineering programs: 68% of the students chose a nursing program and 32% an engineering program. Thus, descriptions based on formulations used in nursing programs were more attractive even for boys than descriptions based on engineering programs, when references to the subject matter were removed.

What nursing programs had and engineering programs lacked

The most popular nursing programme (which was number one in the brochure presented to the students) opened with the paragraph:

The program is right for you if you are outgoing and find it stimulating to meet people, and if you wish to have freedom of choice in regard to your future workplace. You will be able to work with information, supervision, planning and leadership, as well as research and development. Most probably you will work in a team with responsibility for your own area. You will be able to work in Sweden, in the EU or anywhere in the world.

A large number of the students, both girls and boys, perceived themselves as outgoing and chose this nursing education for that reason. The descriptions of the engineering programs did not address this identity at all. (However, there were also students who did not identify themselves as outgoing and were therefore not interested in a nursing program.)

The promise of personal development offered by the nursing programs was also mentioned by a number of the girls, when commenting on the descriptions. In the questionnaire, where 37 different statements were rated by the students, the statement "the education develops you as a person" was the issue that was most important for the girls when choosing an education and even among boys' answers it came high up, ranking number five. There is a total absence of references to personal development in engineering texts. (For boys, the most important issue was the possibility to get a job after the education – which was number two for the girls.)

Nursing programs generally advertise the possibility to spend one term abroad and many of them list the countries with which they have cooperation. Engineering programs almost never do this, even if all Swedish universities have exchange programs for students. The possibility to study abroad was mentioned as one of the main reasons for choosing a nursing program, particularly by the girls. In general, Swedish secondary school students are interested in studying abroad, nationally, more than 60% of the students on the two secondary school programs in this study, girls more often than boys, are interested in doing some of their university studies abroad (Statistics Sweden 2007). The fact that a Swedish nursing certificate is recognized in all EU countries is also often mentioned in program descriptions and many young people, particularly girls, even from these rural communities, obviously are obviously attracted by the possibilities to move abroad to work after graduation.

Nursing texts also often stress the aspects of leadership and independence – presumably because of the long lasting professional struggle to assert the special position and competences of nurses in the medical hierarchy. Some of the students were attracted by these possibilities expressed in the description. However, if these students had known that the leadership and independence were to be executed in a position as a nurse, the words might have had a different flavour.

What were attractive features in the engineering texts?

The first of the engineering programs presented itself as a program with strong ties to working life, with sponsoring companies and even a possibility to have periods of practical work with an ordinary wage integrated in the program. There were even a large number of elective courses. These characteristics were appreciated most by the students who chose the program. The number of electives was appreciated even more than the ties to a workplace, particularly by the girls. Connections to industrial companies, though prominent in the text, were seldom mentioned in the students' motivations for choosing the education. There were even a few students (who did not choose one of the engineering programs) who explicitly did not want to choose an educational program that was connected to industrial companies.

Instead of the connection to companies in itself, the promise that it would result in *real problems* being engaged with in the education was much appreciated. 'Real problems' was one of the code phrases that caught many students. Another such phrase was 'your degree has a heavy weighting in the labour market', which was quoted word by word by several students.

Other code words were *freedom* and *breadth*. While girls more often mentioned 'breadth', boys appreciated 'freedom' more.

'Breadth' is a very frequent word when different engineering programs are described. A broad education is also something that is assumed to interest female students. According to the results of the questionnaire this seems to be true. In traditional descriptions of engineering education 'breadth' denotes a broad technical competence, but today engineering programs assert that even many non-technical subjects and features are included. However, when reading an ordinary recruitment text about an engineering program, it becomes clear that the 'breadth' still primarily concerns different technical areas. It is not sure whether the girls in the study would have reacted as positively to the word, if they had realised that it actually would be limited to a certain area.

The word 'freedom' is not as frequent in program descriptions as 'breadth' and it is interesting to see that it was perceived so positively. However, some of the students (who did not choose an engineering program) also commented on the number of elective courses as something negative, and stated that they would find it difficult to choose among them, because there did not seem to be many guidelines.

Another buzzword in many engineering program presentations is *project work*. This was not at all popular among the students. Only 8 students of the 138 who chose an engineering program stated that project work attracted them to the program, and there were several students who explicitly stated that they did *not* choose an engineering program because of project work. This is particularly true in regard to girls who were even more negative toward project work than boys. Thus, recruiting more girls into engineering is not likely to succeed by increasing the amount of project work.[24]

The possibility to earn money during the program was appreciated, but not as much as the electives, the breadth and the freedom. When it was mentioned, it was often the last item on the list, as an extra bonus.[25]

Those who chose the second engineering program rather than the first did it because it was seen as more *oriented towards people* and as more *practical*. There was a slight difference in how the 'people aspect' in the working life was expressed: The first program mentioned 'co-operation with others' and the second mentioned 'contact with other people, at your workplace as well as all over the world'. None of the students who chose the first program described it as oriented towards people. The phrase in the second description was given by some of the girls as the reason why they chose the second program. Thus, just a slight difference in stressing the 'people aspect' made the second engineering program more attractive than the first for some students.

While the first program described its co-operation with different companies, the second promised profound practical knowledge. This promise of practical knowledge appealed first and foremost to boys. The second engineering program mentioned the degree project and several students commented on it. According to them, it appeared to be too much work. Obvious-

---

24 I presented the results in two classes who had taken part in the study and asked about the dislike for project work. I found out that these students had had lots of ill organized project work during their secondary school studies, and were not interested in having any more.

25 In my presentation in the two classes I also expressed my astonishment that such an offer did not awaken more interest. The students' response was, basically, that they did not expect economic matters to be of primary importance: 'The economic side will always get sorted out in some way'. The offer of paid work might be more attractive for older, returning students, which also are an important target group for three year engineering programs.

ly, the students did not know that all university programs have a degree project, and mentioning it in the description gave the program a negative image. This is an example of how the different frames of reference in the university sphere and among the presumptive students can lead to unintended consequences.

Some more differences between girls' and boys' preferences

The questionnaire about the importance of different aspects when choosing an education was created because all the aspects where descriptions of engineering and nursing education were different, could not be squeezed into the four fake descriptions. Even these statements were fetched from actual descriptions of educational programs in nursing and engineering. They shed further light on what young people of different genders value when choosing an education.

According to these results several aspects promoted in nursing descriptions are more important for girls than for boys, and they mostly have to do with the social aspects of the education, which is not promoted in descriptions of engineering education. Girls more often than boys thought that values and ethics are an important area learn to about, and that it is important to receive training in communication skills. They would appreciate more than boys the efforts of the university to create a good social climate among the students. And girls more than boys thought it an advantage if the education promoted critical thought. Such things are only offered in nursing descriptions. Girls also thought it more important than boys to have an job with 'independence', and this is also offered in nursing descriptions – even if it may be doubted whether nursing jobs actually are more independent than engineering jobs.

Something that emerges from the study is the importance of information about the future labour market. Descriptions of educational programs normally give quite short information about the future profession, compared to the information given about the educational program itself. The fictitious descriptions followed this pattern. However, what was written about the labour market was important for many students in their choice of a program. The possibility to get a job after the education was of primary importance according to the questionnaire, and what was said in the very short description of the future job was often cited as a reason to choose a certain program. It is possible that this weight given on the labour market is, in part, due to the non-academic background of the students. They have probably realised that while secondary education formerly was sufficient for getting a good job, higher education is now required. For them the aim of educating oneself still seems to be to get a good job. And even for these teenagers from the rural Sweden, the possibility to study and work abroad plays an important role.

Discussion

The underlying assumption for this study was that it is not only the subject content that is perceived as gendered in, for example, nursing and engineering education, but also that the way an educational program is organized and described has gendered meanings.

In the first part of the study it was shown that the descriptions of nursing and engineering programs actually conform to societal gender norms: Nursing descriptions stress characteristics that are associated with women and the female dominated sphere of care, and engineering descriptions stress characteristics that are associated with men and the male dominated sphere of production. Thus, it could be expected that descriptions of nursing programmes, even without knowledge of the subject matter, would attract more girls and that, correspondingly, engineering programmes would attract more boys.

The second part of the study showed that this is not necessarily the case. In this sample (with reservations for the possible bias of girls with special interests) nursing attracted only slightly more girls than boys and the fictitious descriptions of engineering education were equally attractive for boys and girls.

The first part of the study puts forward issues about the self-image of the engineering programs. It shows clearly that it is not only technology as a subject matter that makes the image of engineering education mirror different masculine aspects in the society. This can form a background for self-reflection. Why is the focus so heavily on production? Is responsibility really not a characteristic to be promoted in engineers? Is engineering really only about skills and knowledge and not at all about personal characteristics? Is it always advantageous to propose (or require) so much individualism? Looking at a male dominated program in the light of a female dominated program can illuminate the unreflected genderization of the program.

The first conclusion from part two of the study is that even if the descriptions are formulated in a gender-biased manner, this in itself is of minor importance. They attract female as well as male students. The main reason why students make traditionally gendered educational choices is the subject matter. That is, we are still faced with the problem of loosening the ties between technology and masculinity if we want to recruit more girls into engineering. Alternatively, we can try to loosen the ties between engineering and technology (which has, to some extent, been done in Swedish graduate engineering education).

However, paying attention to the formulations in recruitment texts is also of some importance. When choosing between two engineering programs, with similar subject matter, students may well look into aspects that have to do with the organization of studies and the language used to describe them. The descriptions in the catalogues have two target groups: Those secondary school students who still do not know which profession to choose and those who know what they want to study, but who have to be attracted to a particular university. According to this study, it would be difficult to make the students who still are in the process of choosing a profession to make a gender a-typical choice by the design of the educational program and the way it is described in a university catalogue. However, when it comes to those students who already have decided to become engineers, it is possible that the way an educational program is described in a catalogue can play a part in directing them to a certain university. Thus, they may still have some impact in attracting, for example, students of minority gender to a certain university. Using broader descriptions, borrowing expressions even from other discourses than the one prevalent on the educational context, may increase the number of those who get interested in the education.

One reason why program descriptions are stereotypical is probably the competition between the programs. In such a situation, addressing the presumed target group of students similar to those which the program already has managed to recruit is the safest strategy. Engineering educators might be interested in recruiting women, but maybe not on the expense of losing potential male applicants, particularly when these are scarce.

In that light the results of this study are encouraging. Using, for example, formulations more common in descriptions of nursing education, should not 'scare away' male applicants from engineering, as such formulations in the context of the study actually managed to attract more than half of the boys. To further emphasize this point it can be mentioned that, of those 37 students in this study who stated that they actually planned to go into engineering, more than half did choose a nursing program among the four descriptions.

Actually, the students reported very few statements used in the program descriptions making a negative impact. Some students reacted negatively to academic expressions like 'research', 'science', 'specialist knowledge' – which may in part be due to the fact that these students came from environments where adults with higher education are not common. Other students were negative towards project work. However, in general different statements were rated as more or less positive or neutral and few of them would make the impression of the program more negative.

The study has several limitations. Firstly, the students only got two kinds of program descriptions to relate to. If the material had included descriptions from, for example medical education or political science, some aspects that were not included here would have come up in the material and been evaluated by the students in relation to the present aspects and might have proven to be even more important. Secondly, some of the students' opinions have probably been affected by the way the descriptions were written and presented. For example, some aspects that really are not that important may have got extra weight as there only were a restricted number of aspects with which to motivate the choice of the program.

The program descriptions in university catalogues represent, to a large extent[26] the prevailing self-image of the programs. The results of this study should not be regarded as recommendations for changing catalogue descriptions. The fact that young people, even those who plan to study engineering, are more attracted by what nursing programmes advertise, than what is advertised by engineering programs, is a cause for concern for those who work to attract more students, and particularly girls, to engineering. However, rather than reflecting on how texts could be improved, the results call for a more deep going reflection on how engineering programs conceptualise themselves and what they want to provide the young people and their future employers. In some cases the issue is about paying notice to features that already exist on the program but largely go unnoticed (such as the possibility for studies and work abroad), in other cases the issue is about whether a certain feature that does not have an important position on the program would be beneficial for the students and their future employers (such as broader communicative skills, or more attention to students' personal development in general).

The young people in this study were gender transgressing in their choices, within the framework that was provided by the material they responded to. They are still not gender transgressing when it comes to different occupational spheres, such as care or technology. But it is important for the educational programs to be aware that gender norms and preferences seem to become less rigid among the applicants and consider it when they plan for their recruitment of future students.

References

Dufva, Sune G. 2004. Kön, lön och karriär. Sjuksköterskeyrkets omvandling under 1900-talet. Acta Wexionensia, 40/2004. Växjö: Växjö university press

Eisenstein, Z. 1979. Capitalist patriarchy and the case for socialist-feminism. New York: Monthly Review Press

Faulkner, Wendy. 2001. The technology question in feminism: A view from feminist technology studies. Women's Studies International Forum 24 (1): 79-95

Faulkner, Wendy. 2007. 'Nuts and bolts and people'. Gender-troubled engineering identities. Social Studies of Science 37 (3): 331-356

Gilligan, C. 1982. In a Different Voice: Psychological Theory and Women's Development. Cambridge, MA: Harvard University Press

---

26 This article has not dealt with the fact that program descriptions are also affected by the opinions of the PR department of the university. From of the large differences between engineering and nursing descriptions, it can be concluded that the PR departments probably have an influence, but not a decisive one.

The Higher Education Ordinance, In: http://www.hsv.se/lawsandregulations/ thehighereducationordinance.4.5161b99123700c42b07ffe3981.html [3.12.2009]
Högskoleverkets årsrapport 2008, In: http://www.hsv.se/download/18.8f0e4c9119e 2b4a60c800 023245/0819R.pdf [3.12.2009]
Jang, Seongkeun & Yoon, Yongki Lee, Inseong Kim, Jinwoo. 2009. Design-oriented new product development. Research-Technology Management, 52 (2): 36-46
Kelan, Elizabeth K. 2008. Emotions in a Rational Profession: The Gendering of Skills in ICT Work. Gender, Work and Organization. 15 (1): 49-71
Lagesen, Vivian Anette. 2007. The strength of numbers. Social studies of science 37 (1): 67-92
Mulinari, Diana & Sandell, Kerstin. 2009. Location of Gender. A Feminist Re-reading of Theories of Late Modernity: Beck, Giddens and the Location of Gender. Critical Sociology 35 (4): 493-507
Perelman, Leslie C. 1999. The two rhetorics: Design and interpretation in engineering and humanistic discourse. Language and Learning Across the Disciplines, 3 (2): 64-82
Peterson, Helen. 2008. Man måste sälja sig själv: yrkesmässiga krav i det nya arbetslivet: ett könsperspektiv. Växjö: Institutionen för samhällsvetenskap, Växjö universitet
Plakoyiannaki, Emmanuella & (Mathioudaki, Kalliopi & Dimitratos, Pavlos & Zotos, Yorgos. 2008. Images of Women in Online Advertisements of Global Products: Does Sexism Exist? Journal of Business Ethics 83 (1): 101-112
Rommes, Els & Overbeek, Geertjan & Scholte, Ron & Engels, Rutger de Kemp, Raymond. 2007. 'I'm not interested in computers'. Gender-based occupational choices of adolescents. Information, Communication & Society 10 (3): 299-319
Sagebiel, Felicitas & Dahmen, Jennifer. 2006. Masculinities in organizational cultures in engineering education in Europe: results of the European Union project WomEng. European Journal of Engineering education 31, 1: 5-14
Salminen-Karlsson, Minna (2003). Hur skapas den nya teknikens skapare? Ingenjörsutbildningens mansdominans och de kvinnliga teknologernas villkor, in Vem tillhör tekniken? Kunskap och kön i teknikens värld, edited by B. Berner. Lund: Arkiv: 145-173
Statistics Sweden. 2007. Gymnasieungdomars studieintresse läsåret 2007/08. In: http://www.scb.se/statistik/_publikationer/AA9998_2007T02_BR_A40BR0801.p df [4.12.2009]
Statistics Sweden. 2009a. Programnybörjare mot yrkesexamen läsåret 1993/94–2008/09 efter program och kön. In: http://www.scb.se/Statistik/UF/UF0205/ 2008L09/Web_GN3_ProgramNyb.xls. [4.12.2009]
Statistics Sweden. 2009b. Sökande och antagna till högskoleutbildning på grundnivå och avancerad nivå höstterminen 2009. In: http://www.scb.se/statistik/UF/ UF0206/2009H01/UF0206_2009H01_SM_ UF46SM0901.pdf. [7.12.2009]
Wahlström, Kajsa. 2004. Flickor, pojkar och pedagoger: jämställdhetspedagogik i praktiken. Stockholm: Sveriges utbildningsradio
Wajcman, Judy. 1991. Feminism Confronts Technology. Cambridge: Polity Press
Walby, Sylvia. 1990. Theorizing Patriarchy. Oxford: Basil Blackwell
Wolin, Lori D. 2003. Gender Issues in Advertising – An Oversight Synthesis of Research: 1970-2002. Journal of Advertising Research, 43 (1): 111-129

Young people and nanotechnologies

Ilse Marschalek, Petra Moser and Magdalena Strasser

Abstract

For a future technology like nanotechnology, young people in particular are seen as a central public. The EU seventh framework programme funded project NANOYOU is undertaking a communication and outreach programme in nanotechnologies aimed at European youth. The Centre for Social Innovation – Zentrum für Soziale Innovation (ZSI) was responsible for the initial user requirements analysis to identify young people's knowledge, attitudes, specific values, concerns, expectations concerning nanotechnologies and possible gender differences. Furthermore, ZSI is responsible for the evaluation of the outreach materials. Data analysis of all gathered materials revealed that, besides little knowledge about nanotechnologies, young people, both girls and boys, can deal with the complexity of pros and cons, benefits and risks of upcoming technologies. Although they widely believe in positive developments in the future, they remain sceptical and critical concerning major issues such as privacy, consumer protections, environment and health.

Introduction

Nanotechnology (NT) is considered as a technology that is still in its infancy and will unfold its expected promises in the future. Its development and applications will pose further questions about future development but also enable new future projections and visions that need new forms of orientation beyond traditional schemes. Further stakeholder groups have to be involved. The discourse about nanotechnologies therefore leads to wider discussions about societal future in general (see Davies et.al. 2008). Obviously, society cannot be seen as "the public" for NT in general, so communication and outreach activities have to be targeted to specific segments of society. For a future technology like nanotechnology, young people are especially seen as a central public. Of course young people do not respond to the same kind of activities as adults do. Therefore "[…] *communication experts underline that knowing key-audiences is crucial to attain them effectively […]*" (Bonazzi 2010:7). Any outreach and communication activity has to be tailored to the specific needs and interests of the different target groups. Young people are seen as an important target group in future technology discussions because of two main reasons. Firstly, they are considered to be a critical public who accept and adopt new technologies, and they will also be consumers of new products. Secondly, they are considered as future engineers and scientists in the various sectors and fields of nanotechnologies. As the moral and cognitive consciousness of young people has already evolved, a critical discussion about pros and cons of emerging technologies and their social impact can take place with young people. The implemented project activities therefore not only attempt to get an idea about young people's knowledge about nanotechnologies but predominantly about their attitudes and concerns.

The EU seventh framework programme funded project NANOYOU (www.nanoyou.eu) is undertaking a communication and outreach programme in nanotechnologies aimed at European youth. While some FP6 programmes have already put much effort into informing the public about nanotechnologies, they have not focused on youth in particular. The Centre for Social innovation – as one of the nine project partners, including science centres in four European countries – was responsible for the initial user requirements work package as well as for the evaluation of the developed outreaching materials. The requirement analysis, which this paper refers to, included a comprehensive qualitative and quantitative survey to identify young people's knowledge, attitudes, specific values, concerns and expectations concerning nanotechnologies in general, and in the three sub-areas – information and communication technologies, energy and environment, and medicine and health in more detail. Special consideration was given to ELSA – **E**thical, **L**egal and **S**ocial **A**spects of nanotechnologies and how to weigh potential benefits against costs and risks.

The NANOYOU project activities in general target young people between 11 and 25 years of age. As this group covers a very broad age range, it was split into three subgroups of young people between 11 and 14, between 15 and 18 and between 19 and 25 years of age. Additionally, the survey was especially focussing on finding out female and male ways of thinking, as well as interests and attitudes to find out if there are gender specific differences.

This paper outlines the results of the requirement analysis of the work package 1 of the NANOYOU project. It aims at contributing to the discussion about images of science among young people in consideration of gender perspectives.

Methodology

The research methodology consisted of a multi-methodological approach in order to fulfil project aims as well as deliver comprehensive results. The concept of mixing methods can contribute to an elimination of potential weaknesses of a certain method and is often adapted.

Literature review

An initial desk research was carried out about available online material focusing on youth, articles in journals related to science communication of nanotechnologies to the public in general and to young people in particular, ongoing and concluded research projects and undertaken communication activities in different countries.

Expert interviews

Fifteen expert interviews in five countries (Austria, Great Britain, Denmark, Spain and France) were undertaken, both to explore the debate on nanotechnologies in each country as well as to find out about the possible knowledge of young people. Following an interview guideline the interviews explored pros and cons in public debate on nanotechnologies in each country. Furthermore, the interviews were intended to deliver ideas about how youth is confronted with information on nanotechnologies, especially in schools (which schools, which subjects, which stage, obligatory/mandatory, training for teachers, curriculum and the like), but also in extracurricular institutions, such as science centres. The interviews aimed at finding good examples of nanotechnology applications to be discussed within survey activities, finding useful focal points to develop teaching materials and experienced or expected difficulties and gender differences while communicating nanotechnology to youth. Interview partners were engaged in communicating nanotechnology to youth, they were either coordinators for nanotechnology programmes or events for young people, experts working within governmental or school authorities or teachers.

Focus groups

In the run up to the online questionnaire six focus group sessions were carried out in Austria and the UK with different age groups, split into groups consisting of either boys or girls. The aims of the focus groups were to gather associations of respondents with nanotechnologies, to explore the young people's knowledge of nanotechnologies and to explore pros and cons when it comes to nanotechnology and its applications, especially in the course of the discussion when participants get more and more information. When conducting studies on nanotechnologies Zimmer recommends working interactively in smaller groups (see Zimmer et.al. 2008). Special attention was given to ELSA dilemmas. The discussion on ELSA was induced by specific nanotechnology products because *"Participants found it much easier to understand and think about nanotechnologies when they were given concrete ex-*

amples of the ways in which they might be applied." (Opinionleader 2007:9) Thereby, questions could be raised like: Do participants understand which part of the product/application is related to nanotechnologies? Is the product/application interesting, relevant for their lives ...? Will or would they use the products? Is there a threat factor? What would the benefit be? Which ELSA aspect do they mention related to the products?

Online questionnaire

Based upon results of focus groups, expert interviews and existing instruments for surveys dealing with nanotechnologies in this age group, a questionnaire for an online survey was developed. From October 2009 until the end of November 2009 it was available online in eight languages and accessible for all young people within the defined age group (www.nanoyou.eu/survey). The aim of the online survey was to gather quantitative data on young people's opinions, views and knowledge of nanotechnologies. Knowledge about nanotechnology was tested in the form of quiz, solutions were given at the end of the questionnaire. Additionally participants could get some additional information about nanotechnology in the selected fields (energy and environment, medicine, ICT) during the survey.

National context survey

To help with framing the information and data that were collected through the selected qualitative and quantitative approaches, a brief national context report shall give information about the position of nanotechnologies within the national school curricula. In other words, the National Context Survey was designed to collect relevant data and information on the school and educational system. A short questionnaire was sent to experts in school authorities (ministries etc.), who are responsible for or have broad knowledge about the curricula. This contextual framework helps to explain variation in the outcome variables at a number of levels.

Results

Results given here are a summary of the qualitative and quantitative data analysis. The sample size of the online questionnaire after data cleaning[27] was 1.969 as the table 1 shows.

---

27 People with 20 valid answers or less were defined as "drop out" and were excluded from further analysis. Furthermore, people aged 26 plus, were kept for the analysis but results were shown separately.

Table 1: Sample size of online questionnaire

Age group	Total number
11-14	365
15-19	1.302
20-25	105
26 and older	197
Total	**1.969**

Source: Own table

The distribution according to gender of the online questionnaire was relatively balanced. 696 participants were female and 756 were male[28].

Results at hand are based on all cases but additionally those countries that had at least 100 cases were analysed separately. This was the case for Austria, Italy, Romania and Spain. All other countries comprise the category "other countries".

As the sample is not representative, we cannot draw conclusions from sample results about the population of young Europeans. But we can and will use results in an explorative sense: We will get first ideas about some young people's knowledge and attitudes towards nanotechnologies, and the ones we are talking about, are the ones who are very likely to be the most interested, most informed ones or the ones who have the most interested and most informed teachers, friends or other reference persons who made them participate in the survey.

Knowledge and interests

Before answering the question what young people know about Nanotechnology (NT), it is important to point out that most of the young people are familiar with the term 'nanotechnologies'. Similar results were found in other studies on adults (see Hanssen, Walhout, et.al. 2008). However, the actual knowledge about NT is relatively limited. Although many young people have an idea about what NT is, they had difficulties in explaining this complex issue. Some girls of the age group 11-14 had never heard the term NT before. None of the young people could give a definition of NT but most of them knew products made through NT. So, when discussing about the youths' knowledge about NT, it is actually a discussion about products that result from NT, as the following quote shows: '*you do not actually talk about nanotechnologies but rather about products and their characteristics [...] most people are not aware that this is nanotechnology*' (FG male 17-20).

It must be emphasized that the focus group session with the oldest participants, the boys between 17 and 20 years of age, knew the most products,

---

28 There are some samples with missing values on this item.

especially products processed with the nano-sealing. Some of the participants said that the first time they had heard about NT was in relation to nano-sealing and window panes (FG male 17-20). Nevertheless, participants from the younger age group (11-14 years) also knew some nano-products. Besides nano-sealing, which was mentioned often, they had heard about clothes which were treated by NT, sun blockers, etc. On the one hand, it seems that the young people know products but on the other hand, it is difficult for them to understand the NT components of some products. Especially the youngest age group 11-14 could not understand what the NT components of specific products (e.g. socks with silver particles) were, independent of gender. This was also confirmed by other surveys: There is little knowledge about nano-technologies as such but some applications are known. So, for instance, young people know most about self-cleaning surfaces, followed by medical and health applications, according to a German/Swiss study. (Grobe 2007:16)

The results of the online questionnaire show some differences in 'knowledge" about nanotechnologies according to gender. Girls and young women knew less about nanotechnology than boys and young men.

Table 2: Quiz results by gender and age group – all participants that have ever heard of nanotechnology

Number of correct answers	female Age 11 to 14	male Age 11 to 14	Female Age 15 to 25	Male Age 15 to 25
0	3%		1%	
1	9%	2%	2%	1%
2	19%	9%	7%	3%
3	27%	27%	10%	8%
4	18%	23%	20%	13%
5	12%	18%	18%	21%
6	9%	13%	20%	23%
7	4%	8%	13%	19%
8	0	0	8%	10%
9	0	0	2%	3%
	100%	100%	100%	100%
n	68	90	298	479

Source: NANOYOU Online Survey 2009

Differences between females and males are statistically significant in both age groups, T-test for unpaired samples, assumption of non equal variances, alpha = 0,05

Among the 11 to 14 year olds, 31% of girls have only up to two correct answers whereas this percentage is much lower among boys (11%). 25% of girls have 5 or more correct answers, whereas 39% of boys reach these top scores. The same picture holds for the older group.

The difference in correct answers cannot completely be transferred to incorrect answers because females have significantly more "don't know" answers. One could assume that males that did not know the answer just made a guess that sometimes resulted in a correct answer. But even if this assumption holds, it explains only a small part of the differences between females and males. It is a fact: Boys and young men simply know more about nanotechnology.

But the following table gives some hints how these differences are to be reduced: One has to find the context that makes girls interested in nanotechnologies:

Table 3: Curiosity according to fields of application – would you like to give some examples in…?

	female 11 to 14	male 11 to 14	female 15 to 25	Male 15 to 25
Energy and environment	20%	21%	13%	12%
Medicine	29%	13%	28%	13%
Information and communication technologies (ICT)	12%	25%	10%	25%
All three fields	30%	29%	42%	40%
None of them	10%	12%	8%	9%
	100%	100%	100%	100%

Source: NANOYOU Online Survey 2009

Differences between females and males statistically significant in both age groups, chi-square test, alpha=0,05

Females are more interested in medical applications of nanotechnology whereas males are more interested in ICT applications. Interest for energy and environment or all three fields is similar. This pattern holds for both age groups, but at different levels: In the older age group single interest for energy and environment draws back in favour of interest in all three fields.

Concerning gender differences some experts believe that males are often more interested in natural sciences than females. One expert from Denmark reflects on the issue and states:

> "It's more accepted that men are interested. I find that most of the time the girls are a little scared of it, they are scared of math and science. That can really stop them. But then, when we start to talk about it, it normally shows that they know equally as much about it as their fellow students. But I think this is a bigger issue that's based on how the society sees things and not really how we teach things".

In this context a female focus group participant, who was 12 years old, said that she would like to have more information on NT but not in school because of the male classmates, who often laugh about her. Another girl brought in that it is important to foster girls in a certain way because they are basically less encouraged in terms of technical issues than boys. Statements like this could contribute to a new discussion on coeducation.

Attitudes and expectations

In general, young people think that NT will improve our lives in future. They are mostly optimistic up to euphoric but at the same time believe in risks and are aware of negative impacts as well.

For some of them nanotechnology products open infinite opportunities and future possibilities. Young people are aware of its big potential but they also have considerations and ask for information and control.

The findings point out that the information given to young people about NT should also address possible risks and problems.

Young people perceive technology development, which also mirrors the development of nanotechnology, as irresistible and therefore emphasized the positive aspects of it. The development of technology is experienced as a „*law of progress*". Technological progress is almost considered as a natural law, which one has to accept and handle. It seems that young people anticipate its importance in the future.

Young people's seemingly uncritical attitude regarding nanotechnology is caused by a lack of information on possible risks. When youths are informed about those they do show a lot of concern.

All four products that were introduced in the questionnaire were seen as good things by a majority of participants, and best results are given for the sunglasses and the lab-on-a-chip. Interestingly, although opinions on the products are rather positive, a relative majority of young people would not want to have the jacket and the lab-on-a-chip[29].

There is no tendency that the elder group is more critical than the younger group.

For discussing ELSA with young people, concrete examples of NT applications appeared as a successful approach, whereas finding an appropriate medicine example is still challenging.

It is remarkable that girls and young women mentioned concerns and fears more often than their male peers during the focus groups. The fears were not only expressed towards environmental issues but also on health issues and on technical control over human beings. The boys and young men seem to have a more pragmatic approach. In their opinion, instead of referring to the past and always "*clinging to the past*" (focus group males), people

---

29 A lab-on-a-chip is a device that is smaller than a credit card, but can perform all the analysis required to do a blood test. Instead of bringing samples to the laboratory, doctors will be able to use these in their office. In the device there are hundreds of nano-sized detection sites that can recognize a specific "signature" of a disease, such as a specific type of virus.

should be more open minded towards innovation and development. All things have negative and positive aspects and it is not possible to hinder technical development. Relating to the examples of the NT products, one male participant said that *"...you just have to choose the positive things and not get into the negative"* (male, 22 years). One teacher explains that, in his experience, boys are more likely to ignore the risks of certain products whereas girls are a lot more careful. He believes that boys should be told about the worst-case scenario to sensitize them to the possibility of risks regarding nanotechnologies and that less drastic explanations will have the same effect on girls.

Wish for control

In terms of control, young people mostly share the same opinions and attitudes according to the preliminary results of the survey. Control authorities and institutions were claimed. Their internationality and independence were emphasized by most of the young people. Governmental influence instead of private and economic interests behind control mechanism should be assured. Through the involvement of several control authorities young people want to avoid misuse of power in order to achieve a non-hierarchical control mechanism. Besides control mechanisms the aspect of labelling NT products was an issue. Young people want to have information about products which are manufactured by NT, especially in the medical sector where intervention in human bodies is performed. In general, they want to have a voice and right to say where and how far developments on NT should go.

Although most of the young people emphasized the importance of control mechanisms there were many concerns on its autonomy. Especially the boys and young men were very sceptical about the independence of control authorities and stated that the possibility of misuse and *"corruption"* (focus group males 12-15 years) was high. Another male participant stated: *"where there is power, there's fraud"* (male 22 years). However, few of the young people were against control due to the difficulties with independence. *"I think it is pointless, because first you need monitors, then you need monitors for the monitors, etc. and then it would just get ridiculous"* (male 14 years).

In the focus groups, some NT products were discussed with the participants. When debating the socks with nano-silver particles young people were quite sceptical and said that they would not want to wear them before long term testing declared their harmlessness. On this matter, a 14 year old boy said, he does not want to be used as a *'human guinea pig'* before the producers can assure no harm (FG male 12-15).

In terms of privacy young people were very concerned. The following question about a jacket made of nanomaterials, which contains a global positioning system, was answered quite distinctively:

Table 4: Opinions on benefits and risks of nanotechnology applications and products

Jacket	n	Strongly agree	Agree	Disagree	Strongly disagree	Don't know	Total
I don't always want other people to know where I am	1536	33 %	35 %	18 %	8 %	6 %	100 %

Source: Own table

Experts confirmed an increasing sensitivity against this issue. Within the focus group discussions this aspect outweighed the positive aspects of surveillance like support rescue in dangerous situations. Being "*watched*" by other institutions or people was not the only concern but young people were also afraid of surveillance measures by family members, relatives or teachers.

Future

We also asked about further interests and concepts for their career and future. It was obvious that focus group discussions increased awareness and interest. At the end of the discussion rounds all participants stated that they would feel more aware from then on when it comes to the topic. Especially girls had not considered that such a topic could be of interest for them up to this point.

When asked if they wanted to learn more about it, young people tended to agree because they anticipated its importance in the future: "*Definitely, it's the future!*" (focus group 13-15). "*When I see something in the newspaper about it, maybe I would not just continue skimming, because it is related to the future, you cannot elude it*" (female 20 years). Initially, they would like to be informed at school and get further information via media (predominately TV and internet), but they also want to get more insight through exhibitions and concrete examples of actual research and production activities.

Referring to the results of the online questionnaire the differences between boys and girls concerning educational and career goals were not as big as expected in Austria. About 70% of 11 to 14 year old boys and girls think that nanotechnologies are very complicated and too hard to study. Even if about one half of the 15 to 25 year olds still thinks that nanotechnologies are complicated to study about 35% of them are interested in studying nanotechnologies. Internships in a laboratory are appreciated by both age groups similarly. Especially 11 to 14 year olds could be reached by summer camps during holidays. After all, boys are still more often interested in working in nanotechnologies than girls.

The situation is similar among young people from Spain. 82% of all respondents are interested in an internship in a nano-laboratory and 55% in participating in a summer school. Both activities are appreciated by boys and girls similarly.

About one half of the 15 to 25 year olds think that nanotechnologies are very complicated to study (56%). Nevertheless 59% are thinking of studying a subject related to nanotechnologies and even three quarters think that working in nanotechnologies would be interesting. All in all, the findings point out that Spanish respondents especially prefer practical experience to (theoretically) studying nanotechnologies. As a consequence, the interest in nanotechnologies of Spanish young people could be raised with exercises and practical tasks.

Conclusions

As results show, there is a gender gap concerning interest and knowledge in each age group. Concerning nanotechnologies the gender aspects are seemingly similar to other technologies. Accordingly, girls do not feel interested in nanotechnologies as much as their male peers. But seemingly there are potentials. Either girls show a strong interest for the future and would like to learn more, or they are still uncertain whether the topic could interest them later on. Girls in Romania especially showed a lot of enthusiasm in NT even though they are not as often informed. More 15 to 25 year old girls want to study and work in nanotechnologies than boys. To consider these differences some experts suggest not expecting and dealing with any gender differences whereas others are highlighting the importance of dealing with different interests and needs by gender. According to group discussions it seems that educational aspects with basic information about nanotechnology could be better achieved during classes than dialogue and participatory aspects. For interactive formats, the different tolerance levels and fields of interests by gender should be considered.

Even if the proportion of female respondents to surveys in general (Smith 2008), as well as online surveys in particular is typically higher than the proportion of male respondents, in the NANOYOU survey at hand more males participated. The most important aspect is how to gain their interest for the topic in general but also for their self confidence in particular. Mostly, females in particular would not believe that the issue could be of interest for them.

But although they might know little about the subject in general, it works surprisingly well to discuss ELSA and other impacts of nanotechnologies. Young people are very aware of such aspects in general, and willing to consider them when informed about further implications in their closer environments. They are very critical in terms of practical aspects and meaningfulness of products and developments. Discussions about concrete examples offer possibilities for differentiated perspectives. Role plays using dilemmas for stakeholders with different perspectives could be an appropriate approach. Jarmon and Keating believe that "educational role-play scenarios with the active participation of the public can serve as a dynamic method for civic engagement across a range of complex, interdisciplinary topics and new technological dilemmas" (Jarmon & Keating 2008:282). But examples and dilemmas might be different for females and males to cover different fields of interest.

Raising young people's interest in nanotechnologies should be accompanied by information about the relevance for their future. This is a crucial point especially when considering the lower social or educational encouragement of females. Strengthening young women's interest in technology and especially in nanotechnology could work by relating the communication on nanotechnologies to their daily lives and interests so that their views and perspectives for their future will not be limited.

Acknowledgements

Special thanks go to our colleagues from our collaborating partner, the SORA Institute in Vienna and to the project partners of NANOYOU.

References

Bonazzi, Matteo. 2010. Knowledge, Attitudes and Opinions on Nanotechnology across European Youth – Analysis from a specific survey carried out in 25 EU countries. EC/DG Research Directorate G – Industrial Technologies. Unit G.4 – Nano-and converging Sciences and Technologies

Davies, S.; Kearnes, M. & Macnaghten, P. 2008. Nanotechnology and public engagement: a new kind of (social) science? Retrieved March 5, 2009, from javascript:__doPostBack('dnn$ctr3070$Document$grdDocuments$_ctl4$lnkDownload',")

Grobe, A. 2007. "Die Zukunft der Nanotechnologie – Erwartungen von Konsumenten und Experten." Apunto 12/1 2007/08Dr Antje Grobe, Leiterin Nanotechnologie, Stiftung Risiko-Dialog: 16-17

Hanssen, L.; Walhout, B. & van Est, R. 2008. Ten lessons for a nanodialogue: Rathenau Institute

Jarmon, L.; Keating, E. & Toprac, P. 2007. Examining the societal impacts of nanotechnology through simulation: NANO SCENARIO. Simulation & Gaming, 39, 2: 168-181

Opinionleader. 2007. Which? Report on the Citizen's Panel examining nanotechnologies: Opinion Leader

Smith, William G. 2008. Does Gender Influence Online Survey Participation? A Record-linkage Analysis of University Faculty Online Survey Response Behaviour. PhD San José State University, In: http://www.eric.ed.gov:80/ERICDocs/ data/ericdocs2sql/content_storage_01/0000019b/80/3e/06/1f.pdf

Zimmer, R.; Domasch, S.; Scholl, G.; Zschiesche, M.; Petschow, U.; Hertel, R. F. et al. 2007. Nanotechnologien im öffentlichen Diskurs. Deutsche Verbraucherkonferenz mit Votum. Technikfolgenabschätzung – Theorie und Praxis, Nr. 3, 16. Jahrgang: 98-101

Part 2 – Peers, school and media

Besides their parents, peers, school and media[30] can be seen as main influencing instances for young people in occupational or study decision making processes. Although the awareness about the impact of these socialisation agents exists, for instance in youth relevant media meaningful and gender equal SET representations are still seldom to find. It will be discussed how the popularity and image of SET is co-shaped by each of the three instances, in what ways and to what extent they are interconnected. Which influence do they have for young people in decision making processes for or against SET? Last the role of science education in weakening or strengthening of a positive image of SET is discussed.

---

30   Media as socialisation agent will not be focussed in a special article, but results are integrated in Sagebiel et al. (2009), which summarised the results of MOTIVATION project and will be reprinted here at the end of the volume.

Does science education at school make young Europeans like SET?

Anne-Sophie Godfroy

Abstract

The paper is based on some results of the MOTIVATION project, a study funded by the European Union under FP7. The aim of the MOTIVATION project was an exchange between partner countries in Europe about factors, which influence the image of science and technology under gender perspectives to attract young people. The paper explores how a gendered image of science, engineering and technology (SET) is constructed at school and whether it creates an attractive image or not, in the different European countries. It will identify different patterns of science education and their attractiveness towards boys and girls. Secondly, perceptions of pupils will be examined.

As a conclusion, we propose some recommendations and we stress the fact that liking science does not automatically lead to choosing science, even if this assumption lays behind many policy measures. The impacts of the contextualisation of science teaching and the status of science as a discipline play a major role in images and choices.

Introduction

The paper is based on some results of the MOTIVATION project, a study funded by the European Union under FP7. Aim of the project MOTIVATION was an exchange between partner countries in Europe about factors, which influence the image of science and technology under gender perspectives to attract young people. Seven countries participated in the study (Germany, Austria, France, Slovakia, The Netherlands, Spain, Sweden).

This paper will focus on images of science communicated through education at school.

The paper explores how a gendered image of SET is constructed at school and whether it creates an attractive image or not, in the different European countries. It will identify different patterns of science education and their attractiveness towards boys and girls.

In a second time, it will present some attempts to improve the situation through various measures at school: New teaching methods, specific teaching material, single-sex classes in SET, interdisciplinary and problem-based approaches, etc. Perception of pupils and effectiveness of the measures will be examined.

As a conclusion, we will try to propose an overall picture of the European situation through the identification of different patterns, in order to point out the successes and the pitfalls of policies and measures aimed at promoting SET among youth.

Methodology: How are images of science built at school?

The MOTIVATION project tried to identify images of SET built at school and out of school and to assess their impact on the attractiveness of SET careers. On one hand, pupils' images of SET have been studied through drawings, focus groups and interviews; on the other hand, the possible origins of those images have been explored, at school through school books and curricula and pedagogical approaches analysis, out of school through the analysis of images of SET in the media.

Hypotheses

Even if the education of girls has continuously progressed in the last century to equal boys' education and sometimes overpass it, researchers have discovered that far from ensuring equal opportunities for all children, school, unconsciously, creates exclusion through images, habits, language, learning material, teachers' attitudes, etc. Those issues are especially difficult to address because most of the actors are not aware of them.

Marie Duru-Bellat (2004 and before) and Claude Zaidman (1992) have demonstrated that gender differences are built at school in different ways:

- Through peer relations among the children.
- Through attitudes of teachers towards children, often unconsciously gendered.
- Through learning material, representing sometimes only women or addressing only men issues, as if women never existed, or presenting situations where women are always mothers, nurses or saleswomen in a commercial shop, when men are directors, astronauts, engineers, medicine doctors, etc.
- A last effect of gender segregation is produced by the institution itself, which tends to promote men more than women in management positions.

Another major consequence of gendered images of SET is the study and career choice made by pupils. For this reason it is extremely important to raise awareness among teachers and job advisors.

At the same time, school can be a good place to work on gender stereotypes and SET through pluridisciplinary projects, for example.

The paper will study those aspects: How a gendered image of SET is constructed at school and could be challenged at school too by appropriate measures or programmes.

Methodology

Document analysis of school books in SET and general information about school systems have been provided.

In addition to a literature review, documentation on schools systems and collection of national and local statistical data about study choices during secondary and higher education has been compared. For this paper, the study is mostly based on fieldwork: Interviews with pupils and teachers and focus groups.

In each country two different schools were chosen as case studies. The criterion for choice was: "What is "pertinent" in your country?" with a recommendation to choose very different situations, if possible, one in an upper level school with a privileged social background, the other one in a school experiencing difficulties to motivates pupils with a non-privileged social background.

The fieldwork used in this paper consisted in interviews with pupils between 15 and 18, interviews with teachers, and raising awareness workshops. Fieldwork has been conducted in national languages in each country, then analysed and partly translated into English to allow cross-comparison.

In all countries, 25 interviews with teachers, 77 interviews with pupils and 3 raising awareness workshops have been conducted for that study.

Positive and negative images created by teachers

From the answers to some open questions, we can analyse how teachers describe the "dream situation" for SET teaching.

Dream situation for teachers

They would prefer smaller groups (around 15), better material (especially in the "not-privileged" schools) and more time to explain. All teachers stressed the need for more time to support personal work, creativity and autonomy, more time to do more experiments and group-work.

Even though they try to apply those principles in their everyday teaching, they consider that it is not possible with all topics. They favour inclusion and active learning as far as possible. Some teachers prefer various approaches.

"Well, smaller groups would be my dream" (GMT1_4, Germany)

"project-oriented lessons, this would be the third aspect to dream about, having more time for project oriented teaching and planning" (GMT1_3, Germany)

"…maybe a class where I could get my pupils understanding and reasoning and creating to the maximum, and that they were happy with it… but right now I don't know how I could (she laughs)" (ES-01-FTT, Spain)

We can observe some differences from one case study to another:

- In the "privileged" case study: The teachers' focus in on mixing the groups, creating a good atmosphere, developing interest out of the classroom.
- In the "not privileged" case study, they are more concerned with the motivation of students through experiments, having fun, doing things to get their interest.

Gender awareness of teachers is very poor. When asked, most of them answered that they make no difference between boys and girls and that there is nothing special to notice, except maybe the fact that girls feel less confident than boys. Only in the raising awareness workshops did we interview teachers who had thought about gender issues and were aware of it. We will go back to that issue in the conclusion.

Preferred situation by pupils

Dream situations for teachers match the dream situation for pupils. When asked their preferred teaching situation, the pupils answer that it is experiments and hands-on situations. Theoretical work is obviously less appreciated; most of them like doing experiments, and have "fun" with science. Attitudes also depend on the age of pupils (see below).

Some pupils want to think by themselves, some want more guidance; there is no consensus among them.

The pupils' suggestions for improvement are to privilege problem-based learning because it would be less "boring", they would also appreciate "better explanations", better attention to the personal needs, and more connections to everyday life.

"When we do something concrete, it really interest me. When it's really pure theory, like something in maths, I don't like it at all." (FF1_2, France)

The SET teachers satisfy many of them. They compare teachers between each other. They are qualified "good" or "bad", there is no average appreciation:

"The one we have now in physics and mathematics, she is really good, she explains why and how" (SWFI 1, Sweden)

"In technique it was better, because we build things there and the (female) teacher was very nice." (GM1_1, Germany)

"Our chemistry teacher only knows what is written on the sheets she distributes" (GF_2, Germany). [The pupil prefers when pupils bring their own ideas and teachers are able to explain.]

"The teacher really gave me the will to learn this year, she gives us examples, we watch videos, asks for questions: It's really reactive!" (FF2_2, France) [Pupil talking about her biology teacher]

The image of the discipline is always strongly attached to the personality of the teachers:

If the teacher is "good", pupils will like the discipline, if not, it is deterring. This parameter is even stronger than the fact they have good marks or not.

When we ask to pupils to describe a "good" teacher, they answer: Not boring, active, allowing people to ask questions, even out of the programme. They insist on the need to create a "good atmosphere" where the lecture is interesting and everybody is listening.

What is a "bad" teacher? He or she is somebody boring whom nobody listens to, somebody who sticks to "dry" and "dead" approaches that do not trigger the interest.

"My physics teacher is very ambitious, and thanks to her, we can go further than our classes, to understand the mechanism, not only to apply stupidly formulas. It changes everything!" (FM1_1, France) [pupil in Science major].

In our sample we had pupils of different ages and they told us, in some cases, about their opinion a few years ago. The image of SET is obviously changing with age; it corresponds also to changes in the curriculum and the pedagogical approaches related to pupils' age.

Around 10-12 years old, they all share good memories with science experiments (hands-on approaches), when it was proposed in their school. In all cases, it created a positive image of SET and influenced them positively towards science. It was "fun", "not boring", "doing things". In many cases, natural sciences were taught as a whole, as "science", without separate subjects as physics, chemistry, biology, and without mathematical modelling or calculations.

All pupils considered this approach to be very attractive, but some of them found difficult the switch to «serious sciences» when requirements became more complicated some years after.

This kind of approach is sometimes pursued after the early years of high school in high schools that face difficult social contexts. It is the only way to create a minimal attention to science lectures.

Around 14, some pupils are more and more interested. The influence of teacher personality is very strong at that stage. But even at 14, some pupils find SET too hard and boring:

"Initially I found it interesting, but now physics is much more complicated" (GF2_1, Germany)

"Here were million of things to learn by heart (Physics)... In daily life, it has no interest ... S [major in Science at high school], you work a lot, you have a lot of homework, you get sad on Saturday night, you do maths, maths and maths." (FF1_1, France)

In privileged high schools, some pupils have extra lessons to catch up, paid by their parents, or "science clubs" organised by the school. Of course, in schools with a less privileged social background, parents cannot afford to pay for extra lessons, so the difficulties appear earlier.

In the last years at high school (16-18), the differences in SET ability are obvious and in some countries correspond to defined profiles or levels.

In the pupils' mind, there is a clear hierarchy in France between S (science), ES (social sciences and economy) and L (humanities) profiles. S is considered as the most prestigious and difficult, ES is a second choice; L is when you are not accepted in the two first profiles. Of course some pupils will choose profiles because of their taste for a discipline, but the big majority choose according to an implicit hierarchy. Teachers recognize this situation has spoiled personal choice and becomes an obstacle to authentic SET calling.

"I choose Literature as a major when I saw that I had really bad grades in maths. I see the work that they expect you to do in S [science major] and this is not what I want to do...everyone wants to attend S at school. Because this is where you can, as they say, have more open doors, and a wider possibility of jobs. L [literature] would attend preparatory classes, and would become failed writers! And would earn no money! S would rule the world!" (FF1_1, France) [Student who chose a major in literature]

"There is a piloting by the studies you can do, not the careers! Pupils think short-term. They have understood that they will attend some kind of studies and after, they'll see. (...) We have to demolish S. S would only gather good students in sciences. S is now an obstacle to the development of scientific careers. The problem of the S section is that you have to be good everywhere." (FMT2_1, France) [teacher]

The part of theory compared to experiments is increasing in the last years of high school. In Germany for example, a teacher estimated that pupils switch from 1/3 theory and 2/3 experiments to 2/3 theory, 1/3 experiments:

"In the 13th grade the theoretical part increases the closer graduation gets". (GMT2_4, Germany)

Facing this challenge, pupils adopt different attitudes: Some catch up because they have to, they consider that it is the compulsory open gate to something else they like more. Some really like SET and are interested and curious about SET subjects. Some give up because they feel too weak. Among those who abandon, some regret it, some not. There would be something to do for those who have not lost their taste for SET and would be ready for another curriculum.

To finish with pupils' images, their hit parade of SET disciplines is the same in all countries: Biology is first, Chemistry is second, and Physics is third.

How can we explain this result? Recurrent factors in the pupils' discourse are:

- Connections to real life, students like it. It explains biology as number 1.
- More experiments made by students themselves.
- Proportion of mathematics and theory. The more mathematics and theory, the less liked the subject.

We do not mention mathematics and technology because images differ too much from one place to another. Technology does not exist everywhere and corresponds to different topics. Attitudes towards technology do not meet any consensus.

Regarding gendered images, there is not much difference between boys and girls in many cases. Pupils from non-privileged schools seem more positive with SET and especially with technology. Pupils from privileged schools are less positive, and girls seem to give up easier if they experience difficulties and declare they do not like SET. They seem less confident in their SET abilities than boys. We will return to this issue in the conclusions.

Positive and negative images created by books and institutional organisation

Positive and negative images are also created by books and teaching material.

For the books, the approaches differ according to the age and the pupils and according to their profile: Specialised in science or not. Prepared for university or higher education or not.

We can identify two presentations, assuming that real books are always somewhere in between those two extremes:

- The "Science is easy" presentation: This presentation pays attention to context and shows how science or technology can solve everyday life issues. Science is embedded in a pragmatic context or in history and culture (for social science and humanities profiles). This presentation is often developed for pupils who are supposed to be not very strong in science. In some cases, they like it but have already decided that science is not for them.

- "Science is theoretical and difficult" presentation: This approach is developed for scientific profiles and privileges complex and demanding contents. There is almost no context (or very poor), the approach is very theoretical, and mathematics play a very important role. The presentation is elitist, unappealing and perceived as "dry", "dead"," and "difficult" (except for biology).

Figure 1:

Source: *Mathematics for year 4* (secondary education). Authors: Eulalia Vallés, José Manuel Yábar and Neus Margalef (Editorial Teide 2009)

This schoolbook from Spain (Figure 1) illustrates the first type of presentation: *Mathematics for year 4* (secondary education).

A French teacher summarizes this approach:

"With L pupils, they're not so scientific, you may find another approach. You have to link it with actuality. It's far more complicated than S...Lessons are really interactive now (Power Point tools/videos...)." (FFT2_1, France) [Biology teacher]

Here is an example of the other type of presentation: SET textbook in Slovakia:

Figure 2:

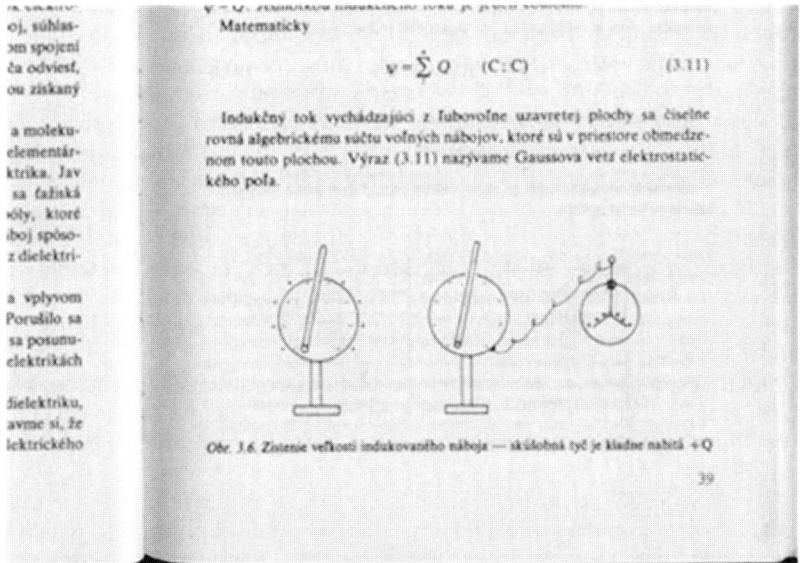

Source: SET textbook in Slovakia

The book is printed in black and white. It has almost no illustrations, only figures. The focus is on a mathematical formula for electricity. It corresponds to what has been described by pupils as disconnected from real life, "dead", "boring", and very theoretical. It seems that context and application is not very highlighted in this book.

Some countries have a common curriculum for all students up to the age of 18, some offer different curriculum from ages 12, 14 or 16.

When there are a variety of offers, the approach of sciences in vocational education, technical education, and general education are not the same. Curricula and approaches may also differ according to profiles "humanities", "social sciences and economics", 'sciences", "sciences & technology", etc.

Generally speaking, vocational education presents sciences embedded in an everyday life context, tries to make them attractive, and avoids theoretical approaches. General education proposes a more theoretical approach, where context is seldom mentioned. Books contain fewer images of persons (except in biology).

In some countries there are some paragraphs about the history of science highlighting famous figures as Archimedes, Newton, Volta, etc. but with very few female figures.

Inside general education, biology is more related to real life than physics or chemistry. Mathematics appears as the "driest" discipline with almost no images, no context, and in some cases colourless books.

When there are different curricula in general education, as in France, these trends are reinforced in the science curriculum, when the humanities curriculum has a very light scientific contents embedded in historical and social context. The result is quite paradoxical: Dry and theoretical approaches are less attractive for pupils, but representations of science and scientists are not obviously gendered (except in the historical chapter). Contextualised approaches are more attractive, but tend to present gendered representations of science: Women are often the patient, the mother, the cashier, etc. while men are engineers, medicine doctors, etc. This was not true of the German sample, where the book presenting SET in everyday life is very sensitive to gender aspects.

In almost all European countries, regulations or laws recommend paying attention to gender issues in school books; in fact, it is not taken into account everywhere for various reasons. One is the lack of money to change books very often.

The challenge would be to have an attractive presentation of sciences (including mathematics) in context, with an attention to gender and without abandoning the substance of the scientific contents. For vocational education, the issues would be a more gender-balanced representation of sciences; it would imply a gender-neutral representation of professions.

Impact on young Europeans: Liking and choosing science

Many projects to support science teaching and promote SET career choice are based on the assumption that liking SET implies choosing SET. In our study, we see that other factors must be taken into account.

Liking is not choosing

There are differences between case 1 (under-privileged social background) and case 2 (privileged social background). The attitude towards technique may differ in the two cases; pupils from case 2 sometimes like technique but prefer school to an. There is a more positive image of technique in case 1. Girls' interest for SET is lower in case 2.

This result presents an analogy with the ROSE project[31] about differences between motivation for SET in over-developed countries and developing countries. This is not surprising if we consider that in some countries, many pupils interviewed in case 1 had a migrant background.

Pupils' dream lessons match with teachers' dream lessons, which is good news. However, there is very poor gender awareness among teachers: Many of them say: "There are no differences" or "I make no differences" just because they are not aware of gender differences. The only exception was in raising awareness workshops, where participants were very conscious of gender differences.

The differences mentioned most often by teachers are:

- Boys are more interested in SET than girls (case 2, not case 1);
- Girls are less confident in their skills.

For pupils, the influence of teachers is huge. They serve as role models for SET jobs and their teaching skills influence positively or negatively students' perceptions of the discipline.

Figure 3

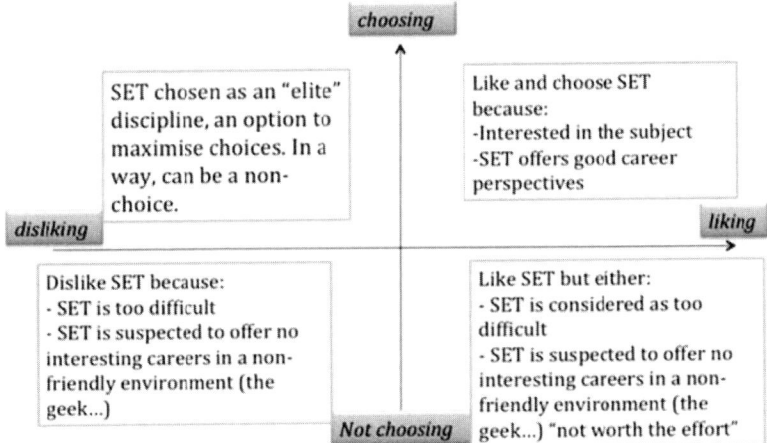

Source: Own figure

We noticed subtle gender effects especially in The Netherlands and in France: There is a gendered balanced proportion of students choosing SET profiles, but sub-profiles in SET are gendered with a high proportion of boys

---

31   http://www.ils.uio.no/english/rose/

in mathematics-physics and a high proportion of girls in biology-chemistry. Regarding France, for example, this situation leads to 26% of women students in engineering, 30% in science preparatory classes, 30% in basic research, 43% in applied research, 62% in biology and 66% in medicine and pharmacy.

As a conclusion, choosing and liking SET are not always correlated. It means that recommendations aiming at making SET attractive and fun are not enough.

Impact of the discipline status

We observed two paradoxical effects of SET images: One is the impact of the discipline status.

In some countries, SET has the image of an elite discipline, mainly because this profile offers many choices for Higher Education. The good point is that it attracts many students for that reason. The bad point is the fact that they choose it as a gate to something else, with no interest in the discipline in many cases.

For strong pupils in SET, even if they choose it at high school, they can abandon it in HE, because they consider that SET are not worth the effort if attractive careers are not there (Slovakia, France). Moreover, it creates a deterring image for average or weak pupils in SET: "too difficult for me".

Sciences, and especially mathematics, are sometimes ways of selecting the best pupils, regardless of their interest for a scientific career later. Science is better considered than economy and social sciences, which are better than humanities. This system has created deep distortions: Science is not a choice or a question of attractiveness, it is a gate to be recognized as having good potential and choice towards any kind of higher education. France exemplifies this situation, which is very damaging for the attractiveness of science in the long term. It attracts good students in the scientific curriculum, but they leave sciences as soon as they can to study business, law, etc. The only exception would be biology and medicine.

On the other hand, when sciences are disciplines as other disciplines, with no special status or even a status of optional discipline, they stimulate a more truthful interest for science.

The challenge would be here to find a good balance between a science curriculum, which is not an option or a secondary discipline, and an elitist approach of sciences aimed at selecting good students, even if it makes them dislike sciences.

From a gender perspective, the results are again a little bit paradoxical: Where what matters is good marks more than interest in science, it may attract pupils who would have never thought to study sciences, especially girls. In fact, the proportion of boys and girls is balanced in the French and Dutch scientific courses, but few of them want to go on in the discipline, except towards medicine. Where it is a matter of interest, we should not underestimate the weight of stereotypes.

Impact of SET in context

Another paradoxical effect is the impact of SET taught in context or not.
When SET are taught in context, they are more attractive, but there is a higher risk of presenting a gendered context not attractive for girls. This contextualisation is mostly used in beginners courses or courses aimed at pupils who do not like SET.

Recommendations for SET at school

As a conclusion of this study, we can propose some recommendations. The essential finding is that pupils' and teachers' motivations are essential, even if liking and choosing SET are not always correlated.

First of all, teachers' training is essential regarding both gender awareness and pedagogy. As principal role models for the pupils and as keyplayers in the transmission of SET images, they have to be trained efficiently regarding pedagogy and gender. Training in pedagogy is essential because perceiving a „good" teacher in SET is the first reason for pupil to be enthusiastic and motivated to develop an interest in SET. In some cases, teachers have to develop self-confidence to become better teachers as they have chosen this position as a second choice because they were not particularly successful students in SET. An effective pedagogical training is particularly necessary for those teachers. Gender must be focused on because gender awareness is very poor, this situation contributes to unconscious stereotypes and less girls motivated and self confident to study SET.

Second, the elitist approach of SET developed in some countries must be questioned. Is this approach a real benefit for the subject? It attracts gifted students into science, but it creates an image of science as a tool for selecting the best potentials. The effect is very often theoretical and difficult courses in order to test students. Many able students are not motivated by such an approach and consider SET as an unpleasant and necessary moment in their curriculum, before being allowed to do something more interesting. For the weak ones, such an elitist approach is completely deterring.

Third, curricula have to be adapted to the XXIst century, many teachers and pupils complain about it. SET contents embedded in a contemporary context would be more interested for students. Science and society issues, controversies, everyday life applications and experiments have to be developed. The example of biology demonstrates that connections to real life and experiments attract students and especially female students.

Fourth, pedagogical research on the transition from "easy" hands-on science to advanced levels should be developed. If the interest for SET is correctly developed in the first years of high school, many pupils feel bored or overwhelmed by SET when more theoretical and mathematical approaches are introduced. How can these be accompanied along with introducing difficult concepts in a more pedagogical way?

References

Baudelot, Christian & Establet, Roger. 1992. Allez les filles! 2nd edition. Paris: Seuil
Duru-Bellat, Marie. 2004. L'école des filles: Quelle formation pour quels rôles sociaux? Paris: L'Harmattan
Mosconi, Nicole & Stevanovic, Biljana. 2007. Genre et Avenir: Les représentations des métiers chez les adolescentes et les adolescents. Paris: L'Harmattan
Mosconi, Nicole. 1998. Egalité des sexes en éducation et formation. Paris: Presses Universitaires de France
Sjøberg, Svein & Schreiner, Camilla (coord.): ROSE project: The Relevance of Science Education. http://www.ils.uio.no/english/rose/
Vitry, Daniel (ed.). 2009. Filles et Garçons sur le chemin de l'Egalité de l'école à l'enseignement supérieur. Paris: Ministère de l'Education Nationale-Ministère de l'Enseignement Supérieur et de la Recherche. http://media.education.gouv.fr/file/2009/33/6/F_&_G_sur_le_chemin_de_l_egalite_2009_web_45336.pdf
Zaidman, Claude. 1996. La mixité à l'école primaire. Paris: L'Harmattan, Coll. Bibliothèque du féminisme

Teach the teachers: Gender competence as an innovative element of teacher-training in mathematics

Anina Mischau, Bettina Langfeldt and Karin Griffiths

Abstract

Regarding the development of skills and interests, results of national and international standardised assessment tests show that pupils are split between the sexes into specific subject areas. In Germany, mathematics can still be considered a typical subject for boys. For more than 15 years mathematics education experts in the field of gender studies have been emphasizing not only the development and realisation of a gender-sensitive mathematics education in schools to be important, but that the sensitisation of mathematics teachers with respect to their role in production and reproduction of gender stereotypes is a necessary precondition. Until now a systematic and comprehensive integration of gender competence in mathematics teacher-training has not been available. In the research project "Gender competence as an innovative element within the professionalisation of teacher-training in mathematics", a gender competence module element will be developed, tested, and evaluated. After completion of the project, it will be available as a learning unit for every university. Besides initial experiences the following paper presents the conception, contents and educational objectives of this gender competence module element.

Introduction

Results from international studies such as PISA or TIMSS have up to today shown gender differences in mathematics achievement and mathematics self-concept. Germany belongs to those countries in which statistically significant gender differences in favour of boys were found. This gender differences already become apparent at the end of primary school and obviously get stronger through the course of schooling. Considering further criteria, for example, mathematical content areas, cognitive achievement levels, types of school, proficiency levels etc., these indicate that gender differences still vary in their characteristics. But overall, mathematics in Germany still seem to be a "typical boys' domain".

The German results in the international comparison studies provoked a new discussion on necessary development and control of quality in the German education system. The initial point of this debate was the question: What is so different in other countries, which achieved noticeably better results than Germany? In search of possible explanations, German scientists analyzed, for example, the influence of the structure in the respective education and school system, the practiced teaching culture or the pedagogical practice in other countries' school systems, the national curriculum framework or the instruments responsible for the regulation and control of quality in schools (cf. BMBF 2001 2007). Sometimes these debates were also connected to questions on the professionalism of teachers, teacher-training and further education (cf. BMBF 2007).

However, a gender perspective is hard to find in all these debates on quality or professionalism. Although for many years gender related school and education research has been discussing the question, whether the structure of the educational system or the prospective teaching culture has any influence on (national) gender inequalities (cf. Grevholm & Hanna 1995; Zimmer et al. 2003). Even the OECD suggested in their publications on PISA 2003 an influence of these factors on the development and reproduction of gender specific "interests and knowledge areas" as well as on school performance (cf. OECD 2004: 99). Further indicators for the explanation of such gender related differences in mathematics performance and mathematics selfconcept could be found in the formation of societal gender stereotypes or more or less gendered disciplines and school subjects (cf. Steinthorsdottir & Sriraman 2008; Zimmer et al. 2003). Furthermore, people within didactics and gender related educational science stated for more than 15 years that besides the development and implementation of gender sensitive didactics there is also the necessity of a sensitisation of maths teachers' contribution to creation and reproduction of gender stereotyped domains (cf. Buchmayer 2008; Jungwirth & Stadler 2005; Keller 1998; Ziegler, Kuhn & Heller 1998). Besides other elements, gender stereotyping and the lack of gender competence on the part of teachers must be identified as a fundamental reason for the subject specific gender bias within the development of interests and competencies of young people. Unfortunately, such discussions have not taken place in mainstream debates on the professionalism of teachers in Germany.

The recently published special report on the PISA analysis picks out this theme in particular as a central issue and discusses gender related disadvantages or stereotypes in our society and on the part of teachers and their influence on the performance of boys and girls in school. Moreover, this report concludes that school as an institution clearly supports girls and boys in different ways and that teachers can contribute to the development of gender equality to a much greater extent, as it has been discussed until now (cf. OECD 2009). Therefore, this special report shares the authors' perspective, which is generally speaking: Gender competence must be seen as an important aspect of the professionalism of teachers and gender sensitive didactics need to be weaved into the development and control of quality in the educational system.

Thus, prospective teachers, in particular, as future multiplicators need to acquire and include gender competence in their didactics and subject specific contents (cf. Keitel-Kreidt 2007), in order to participate in overcoming gender specific connotations within school subjects. Particularly in mathematics and natural sciences, there is a lack of possibilities for student teachers to concentrate on gender topics. A systematic integration of gender competence in maths teacher-training would be a sustainable step towards an increase of equal opportunities, beyond gender stereotyped "knowledge and interests areas". In addition to mediating subject specific contents and methods, a sustainable curriculum reform aims at including gender competence as a supplementary part of the teacher-training. This additional element should not only be subject specific but also general. Thus, it would be an important component within the professionalisation, the strengthening of competence, and the quality assurance of an occupation oriented academic training of maths teachers.

The project "Gender competence as an innovative element within the professionalisation of teacher-training in mathematics" points to the lack of an integration of gender competence within teacher-training in mathematics.[32] This builds an overdue bridge between academic structure reform, gender mainstreaming, and the debate on the professionalisation of teacher-training. Furthermore, this project tackles two deficiencies: The lack of professional experience and the lack of a gender perspective within the mediation of subject specific and didactic contents (cf. Mischau, Langfeldt & Mehlmann 2009). It aims at the development of a gender competence course for the teacher-training in mathematics. This course, titled "Mathematics – School – Gender", was tested during the winter semester 2009/10 at eight German universities. Moreover, it was evaluated and subsequently allocated as a transferable tutorial at other universities. In the following, conception, contents, and educational objectives of this gender competence course and initial experiences will be described.[33]

---

32 The project "GenderMathematik: Genderkompetenz als innovatives Element der LehrerInnenausbildung für das Fach Mathematik" ("GenderMathematics: Gender competence as an innovative element of the professionalisation of teacher training in mathematics") was funded from 2008-2010 by the German Federal Ministry of Education and Research. The project was a collaboration of the universities of Bielefeld, Gießen and Hamburg (see http://www.uni-bielefeld.de/IFF/genderundmathe/index.html).

33 This paper was written in 2010. Meanwhile, based on the evaluation, the initial course concept was revised (cf. Mischau et al. 2010). In particular, the connection to professional practice was highlighted by introducing the distinction between the teaching framework and the classroom interactions (see Langfeldt et al. 2012). To date, the revised course has successfully been taught five times: By project head Anina Mischau, as a visiting professor at Freie Universität Berlin (summer term 2011, winter term 2011/2012 and winter term 2012/13) (cf. Mischau 2012), by Renate Motzer, a cooperation partner at University of

The gender competence course

The basic conception of the course is to convey an understanding of gender competence as an occupational related key qualification, which consists of the following three dimensions:
- Basic and expert knowledge in the field of gender studies. Foremost about socio-cultural construction modi of gender and their impacts on societal structures, institutions, individual acting, but also on the development of scientific disciplines and gendered disciplinary cultures (gender knowledge).
- Methodological and didactical competences with regard to a gender sensitive composition of teaching and learning processes (acting competence).
- Self-reflectivity within the occupational practice (gender related self- and social competence)

The course aims first of all – and this is strongly related to the understanding of gender competence – on a sensitisation of the students for the mechanisms and consequences of gender stereotyped attributions for the reproduction of gendered "interests and knowledge areas" in the fields of mathematics achievement and mathematics self-concept. The second aim is to get to know important theoretical concepts and methodological and didactical approaches for a gender sensitive mathematics lesson. The course intends, thirdly, to teach competences, which later in the classroom situation allow the students to recognize and perform their role as a teacher under a gender perspective. This includes a critical reflection of their own gender related views and positions as well as the specific institutional frameworks and guidelines (e.g. gender hierarchies and gender segregation in school, curricula, educational material). Additionally, abilities to question the own occupational acting with regard to possible gender stereotyping effects are needed. Gender competence is therefore not a "formula" knowledge. It refers, rather, to an understanding of gender related construction processes within school and identifies traps of co-construction by own attributions, by forms of teaching structures, by patterns of interaction etc. It furthermore aims at developing and proving adequate problem solving strategies.

The module contains 13 fully designed sessions. The three main dimensions of gender competence (gender knowledge, methodological and didactical competence, self-reflectivity) are to be found in each session, though with varying extent. The "double-decker" approach was used during the structuring of all sessions, which means: Some of the methodological and didactical

---

Augsburg (winter term 2011/2012) and by Claudia Lack, a cooperation partner at Justus-Liebig-University Gießen (summer term 2012).

considerations that are qualified for a gender sensitive mathematics lesson were integrated into the concept of the module itself. A centre stage is taken primarily by different forms of cooperative learning. The students will get to know and learn to use some of these learning strategies. The themes and questions of each session will be developed mainly through different forms of teamwork and work in pairs, using, for example, the Jigsaw or the Think-Pair-Share strategy. Another focus of the module's learning forms is the trial of diverse presentation methods. To increase the prospective teachers' self-reflectivity the students will work on a portfolio during the whole course. This portfolio is used for documenting and reflecting results, but also experiences, problems and findings of contents and methods of each session. In the following, a brief outline of the conception and contents of each session will be given.

Session 1

The module's central focus is to sensitise the students for the problematic effects of gender stereotyping attributions made by teachers and pupils. These gender stereotyping attributions concern mathematics in general as well as beliefs about mathematical talent and competence. The sensitisation includes the reflection of subjective experiences with mathematics, personal mathematical view of the world and personal (gendered) beliefs about mathematics. The aim of the first session is to trace these attributions. This happens, first of all by presenting pictures about mathematics in order to make the students compare these with their own. The second step focuses on gendering of mathematics, which results, among other things, in gender stereotyped ability attributions. The students shall discover their own ideas and images regarding the question, whether "being good in maths" is probably more "male" or "female" connoted.

Results of research on epistemological beliefs refer to the fact that besides didactical and purely subject related knowledge of teachers, their own mathematical view of the world and ideas of teaching and learning also connected to this view, have an important influence on the organisation of lessons (cf. Blömeke, Kaiser & Lehmann 2008; Grigutsch, Raatz & Turner 1998; Köller, Baumert & Neubrand 2000). Therefore, the intention of the first session is to sensitise the students to realizing that their own pictures of mathematics and their gender stereotyped beliefs have an influence on their prospective teaching practice. So, the central dimension of self-reflectivity is initiated already in the first session and will be continued during the whole course.

Sessions 2 and 3

Sessions 2 and 3 of the course concentrate on gender differences regarding mathematics achievement and mathematics self-concept. Thereby some selected results of international as well as national assessment studies will be presented. The analysis of these results is important for many reasons: Results from national as well as international comparative studies on students' performance point out that school subjects are divided into gender-specific domains regarding the development of pupils' interests, motivation, self-confidence, academic self-concepts and achievements (cf. Kaiser-Meßmer 1993; Kampshoff 2007; Stürzer et al. 2003; vbw 2009). Mass media, in particular, reduce these findings to exaggerated head-lines: "girls can't do math and boys can't read". Such gender stereotyping summarisations ignore the fact, that the results of these studies do not show a uniform and general picture of gender related differences. The studies demonstrate that indeed gender related differences in mathematics achievement and mathematics self-concept still occur and mostly to the disadvantage of girls. Furthermore, differentiated analyses show that gender differences are however more or less developed. For example they vary according to countries, mathematical content areas, cognitive achievement levels, types of school, proficiency levels, levels of education, type of tests, and they can obviously change during school years (cf. Bos et al. 2003 and 2008; Helmke et al. 2002; Lehmann, Rainer H. et al. 2000; Lehmann, Rainer H. et al. 2001; Mullis et al. 2000; OECD 2004 and 2009; Stanat & Kunter 2001).

Consequently both sessions try to change the students' view and make it possible for them to differentiate between gender stereotyped generalisations of empirical results by mass media or other social actors and real empirical findings. Furthermore, they will be sensitised to realize that phrases like "girls can't do maths" sustain the reproduction of gender stereotypes, and that these phrases' absoluteness is empirically not solid. In terms of self-reflectivity, the Sessions 2 and 3 intend to make the students aware of their own gender stereotyping images, which possibly were created with the help of shortened findings of comparative studies pictured by mass media but also by primary and secondary institutions of socialisation.

Sessions 4 and 5

The question for reasons of gender differences in mathematics performance and mathematics self-concept discovered in comparative studies is the main point of Sessions 4 and 5. The students begin in Session 4 with a selection of different approaches of explanation. They range from biological to socialisation theory approaches (cf. Beermann & Heller 1990; Keller 1998;

Tausenpfund 2007). The students will examine theses reasons by the medium of some extracts of interviews with other students, who expressed some of these explanations. This method enables the students to launch a reflection process on their own patterns of explanation shaped by everyday life experiences. Furthermore, it deepens the sensitisation of personal gender stereotyped prejudices or attitudes that are potentially part of these patterns of explanation. Afterwards, the results of the everyday life explanations will be compared with the scientific discussion on the respective approaches of explanation. The critical examination of these approaches on gender differences in mathematics performance and self-concept is crucial for recognizing the problematic role of the gender stereotyping self attribution and attribution by others. The intended self-reflection aims at sensitising the students for this problem. This also includes drawing attention to perception, attitudes, and behavioural expectations of the prospective teachers as an unconscious part of the "hidden curriculum".

Session 5 connects to the socialisation theory approaches presented in Session 4, which direct the view on the contribution of school and the role of teachers for the formation and reproduction of established gender stereotypes. Subject matters of this session are the findings of the gender related school and lesson research on the "hidden curriculum", through which – unconsciously and unintentionally – gender stereotypes are reproduced in school (cf. Nyssen 2004; Stürzer et al. 2003; Valtin 1993). Two central aspects of the multifaceted dimensions of the "hidden curriculum" are gathered as exemplary examples: Patterns of interaction between teachers and pupils and educational material (cf. Jungwirth 1991; Keller 1998; Kriege 1995). These two aspects were chosen because teachers could have a direct influence on them. On the one hand, through a more careful selection of educational material and through an explicit examination of gender stereotypes in schoolbooks; on the other hand, through the reflection of personal stereotypes reproduced by interacting in lessons and through the practice of alternative patterns of action, that avoid these stereotypes. The main task in Session 5 is the analysis of schoolbooks and studies on interaction. The students shall recognize gender stereotyped images in schoolbooks and the role of teachers regarding gender stereotyped interactions during the lesson. Furthermore, the students shall be sensitised for alternative patterns of action.

Sessions 6 and 7

Sessions 6 and 7 provide an insight into the possibilities of alternative courses of action in terms of a gender sensitive mathematics lesson. An initial approach takes place in Session 6, in which comprehensive educational concepts or educational approaches will be discussed. These are, in detail: The concept of a reflexive coeducation of "doing gender" in school as well as the concept of a meaningful mathematics lesson. The concept of a reflexive coeducation aims at "changing [...] the gender ratio to the advantage of an equal cohabit" and it emphasises the "perception and handling of gender stereotypes [...] in all school subjects" (Bildungskommission NRW 1995: 126 and 131). The concept of "doing gender" focuses in contrast more on active processes of social construction of gender in the school and classroom. Thereby, not only were mechanisms shown through which in interactions during the lesson gender stereotyped restrictions and one-sided perceptions were created but also methods by which these mechanisms can be neutralised will be enlightened (Nyssen 2004: 405). Regarding innovative mathematics lessons, the most prominent German approach is the concept of the "meaningful mathematics lesson for boys and girls" (Jahnke-Klein 2001). This concept advocates a multifaceted and open mathematics lesson, and a gender conscious approach on the organisation of lessons in general. The students will learn, with a critical examination of these three concepts, the potentials and barriers of overcoming the "hidden curriculum" in school and within lessons. Besides, it shall be worked out to what extent these approaches open up possibilities and patterns of action for a gender sensitive mathematics lesson.

Subsequently, Session 7 discusses more concretely different methodological and didactical approaches as well as concrete teaching methods that seem to be especially qualified for a gender sensitive mathematics lesson. The spectrum of the selected approaches ranges from didactical principals (individualisation and differentiation), teaching principles (action orientation), didactical (dialogical mathematics lesson) and methodological approaches (cooperative learning), over concrete methods and learning strategies (jigsaw, carousel workshop, weekly schedule workshop), specific access to mathematics (learning by history) to alternative structuring methods (mindmap, cluster and the like) (cf. Barzel, Büchter & Leuders 2007; Brünning & Saum 2006; Druyen 2005; Gallin & Ruf 1998; Hettrich 2005; Huwendiek 2004; Leuders 2003).

The development and reflection of different didactical and methodological approaches as well as instruments is particularly relevant for the students' future occupational practice. The session aims at the acquirement of a differentiated methodological and didactical knowledge, which opens up alternative actions for the planning and the implementation of a gender sensitive mathematics lesson. The knowledge about the possible applications and effects of these different instruments is a crucial factor of gender competence as methodological competence.

From session 8 onwards

The Sessions from Session 8 onwards build even more of a bridge to the students' prospective occupational practice. The aim of this section is to connect the developed knowledge about the reproduction of gender related domains (regarding the development of pupils' interests and knowledge) and the methodological and didactical instruments for a gender sensitive mathematics lesson with concrete mathematical contents. Furthermore, the students shall get the possibility to produce own ideas of a gender sensitive lesson and to concretise these with the help of planning an exemplary lesson in groups. Sessions 8 and 9 serve as a preparation for this exercise. The concrete implementation takes a centre stage from Session 10 onwards.

Selected best-practice examples for good mathematics lessons will be discussed and analysed with reference to gender aspects in Session 8. The students' task is to examine whether these best-practice examples were at the same time gender sensitive. On the one hand the students need to pay attention to the gender sensitivity of the used material and methods. On the other hand, possible difficulties in concrete lesson settings, like interaction, need to be addressed. Furthermore, the students need to question whether these examples were free from gender stereotyping attributions regarding their proposed implementation of mathematical contents within the mathematics lesson. Session 8 aims therefore at a more detailed and concrete implementation of the previously acquired knowledge about the arrangement of a gender sensitive lesson and about gender competent interaction in school context. Furthermore, the critical reflection of the best-practice examples serves as a sensitisation for the possibilities and potential barriers of a gender competent implementation of "good" mathematics lesson. The practice and reflection of the previously acquired knowledge are strongly important fundamentals for the project work beginning with Session 9.

Small groups for the upcoming planning of an exemplary lesson will be formed in Session 9. The students can choose mathematical themes regarding their respective degree programme (primary area, secondary I and secondary II areas). With the now beginning project work the students shall develop an outline of a mathematics lesson. They shall not forget to use gender sensitive methods and open up space for the teachers' gender competent action. Session 10 serves as a supervised and individualized instruction where the facilitator advises the students in small groups. The students get the possibility to reflect their present thoughts and preliminary work with the supervisor, before they finish their exemplary lesson. In the following two (or three) sessions the students get the chance to present their results. They can choose the form of presentation and the results will be afterwards discussed and reflected with all participants within a plenum. All participants judge the results by using a specific standardized "observation form", which will be filled in by the students during every presentation. The students shall evaluate the form and contents of the presentation, emphasise positive aspects and name critical points. The observation forms will be handed out to the project groups, so that they can transfer them as part of the self-reflection into the portfolio. Furthermore, the students will practice a feedback technique that is also necessary in the everyday life of the future occupation.

The final session of the course aims at summarizing the previously acquired knowledge and at discussing open questions. Such questions could for example regard the implementation of the developed approaches for a gender sensitive mathematics lesson. Another important aspect within the final discussion could be the question, if the students see any need for change in the mathematics teacher-training at university with regard to the transfer of gender competence.

Final conclusion

Although the German Rectors' Conference (HRK 2006) and other important institutions of Germany's higher education system recommend a) to implement more praxis-oriented modules into the curricula of teacher-training and b) to convey more key competencies, we experienced some problems with offering our course "Mathematics – School – Gender". First, collecting information about the curriculum reform at all German universities and looking into details of several specific curricula, due to the ongoing process of professionalisation of academic teaching on the one hand and the Bologna Process on the other hand, left the impression that in the view of many decision makers gender competence is not considered a key competence for prospective mathematics teachers. Second, modular coordination is often strongly restricted and not open for additional course offerings. We were most successful in terms of finding a suitable frame for our course (and consequently to get enough students) when in the didactics of mathematics a module called "special aspects of didactics" or anything comparable existed. By contrast, it turned out to be very difficult to place the course into the special branch of mathematics and caused only few responses from students. The third, and in anticipation of our course from experts in group-discussions already mentioned, problem was to find lecturers in mathematics or the didactics of mathematics who are able to a) teach contents with gender aspects and to perform this b) in a gender sensitive manner. Therefore we trained the external lecturers of our project very comprehensively and tried to impart the substantial and didactical concept of every single lesson. Besides need for improvement, the evaluation of the courses at all eight German universities will probably reveal whether we succeeded to communicate our main goals to the lecturers or not.

The first wave of our small panel survey shows that students taking part in the course "Mathematics – School – Gender" judge the actual amount of the didactics of mathematics and the share of educational training in their curriculum as too small. They also mention a lack of didactical and methodological knowledge to deal with gender differences at school. Additionally, about 50% of the students report having little competence and experience in self-reflection and self-observation. The expectations of the students concerning the course "Mathematics – School – Gender" are in line with these first findings of the survey and correspond with the objective of our course-design. Students hope to get informed about recent findings in the field of empirical educational research, they want to learn more about pupils' gender differences in mathematics and how to deal with these differences, respectively, how to avoid the development of gender specific interests. The statements written down in this open question of our questionnaire stress the deficit of praxis-oriented course offers and send two crucial signals. First, many prospective teachers do not feel well prepared for real school life. They think a lot about their role as a teacher but they are uncertain how to cope with internal and external expectations. Second, there is a high interest in differentiated and individual teaching methods. Students know that it is necessary to deal with diversity at their future workplace and many of them correctly regard gender as one main aspect within this context.

References

Barzel, Bärbel; Büchter, Andreas & Leuders, Timo. 2007. Mathematik Methodik. Handbuch für die Sekundarstufe I und II. Cornelsen Verlag Scriptor: Berlin

Beermann, Lilly; Heller, Kurt A. & Menacher, Pauline. 1992. Mathe: nichts für Mädchen? Begabung und Geschlecht am Beispiel Mathematik, Naturwissenschaften und Technik. Huber: Bern

Bildungskommission NRW. 1995. Zukunft der Bildung – Schule der Zukunft. Denkschrift der Kommission „Zukunft der Bildung – Schule der Zukunft" beim Ministerpräsidenten des Landes Nordrhein-Westfalen. Neuwied

Blömeke, Sigrid; Kaiser, Gabriele & Lehmann, Rainer. 2008. Professionelle Kompetenz angehender Lehrerinnen und Lehrer. Waxmann: Münster

BMBF (Bundesministerium für Bildung und Forschung). 2001. TIMSS – Impulse für Schule und Unterricht. Forschungsbefunde, Reforminitiativen, Praxisberichte und Video-Dokumentation. Bonn

BMBF (Bundesministerium für Bildung und Forschung). 2007. Vertiefender Vergleich der Schulsysteme ausgewählter PISA-Teilnehmerstaaten. Bildungsforschung Band 2, 3rd Edition. Bonn

Bos, Wilfried et al. (eds.). 2008. TIMSS 2007. Mathematische und naturwissenschaftliche Kompetenz von Grundschülern in Deutschland im internationalen Vergleich. Waxmann: Münster

Bos, Wilfried et al. (eds.). 2003. Erste Ergebnisse aus IGLU. Schülerleistungen am Ende der vierten Jahrgangsstufe im internationalen Vergleich. Waxmann: Münster

Brüning, Ludger & Saum, Tobias. 2006. Erfolgreich unterrichten durch Kooperatives Lernen. Strategien zur Schüleraktivierung. Mit einem Vorwort von Kathy und Norm Green. 2nd Edition. Neue Deutsche Schule: Essen

Buchmayer, Maria (eds.). 2008. Geschlecht lernen. Gendersensible Didaktik und Pädagogik. Studienverlag: Innsbruck/Wien/Bozen

Druyen, Carmen. 2005. Fünf Basiselemente Kooperativen Lernens. In:. http://www.toolbox-bildung.de/fileadmin/user_upload/Bausteine_Schule/ 044_Kooperatives_Lernen/5_Basiselemente.pdf. [22.01.2013]

Gallin, Peter & Ruf, Urs. 1998. Dialogisches Lernen in Sprache und Mathematik. Band 1: Austausch unter Ungleichen. Grundzüge einer interaktiven und fächerübergreifenden Didaktik. Kallmeyer: Seelze-Velber

Grevholm, Barbro & Hanna, Gila (eds.). 1995. Gender and Mathematics Education. Lund University Press

Grigutsch, Stefan; Raatz, Ulrich & Törner, Günter. 1998. Einstellungen gegenüber Mathematik bei Mathematiklehrern. Journal für Mathematik-Didaktik, 19: 3-45

Helmke, Andreas et al. 2002. Das Projekt MARKUS (Mathematikgesamterhebung Rheinland-Pfalz: Kompetenzen, Unterrichtsmerkmale, Schulkontext) (Abschließender Kurzbericht). Ministerium für Bildung, Frauen und Jugend: Mainz. In: http://www.lars-balzer.info/publications/pub-balzer_2002-01_MARKUS2002-Kurzbericht.pdf [22.01.2013]

Hettrich, Monica. 2005. Entdecken, Erleben, Beschreiben – Der Dialogische Mathematikunterricht – Ein Projekt zur veränderten Unterrichtskultur in Mathematik. In: http://www.dialogischer-mathematikunterricht.de/artikel_magazin_schule_ 15.pdf [22.01.2013]

HRK. 2006. Empfehlung zur Zukunft der Lehrerbildung an Hochschulen. Entschließung des 206. Plenums am 21.02.2006. Bonn

Huwendiek, Volker. 2004. Didaktische Modelle, in Leitfaden Schulpraxis. Pädagogik und Psychologie für den Lehrberuf, edited by Gislinde Bovet & Volker Huwendiek. 4th Edition. Cornelsen Verlag Scriptor: Berlin: 31-67

Jahnke-Klein, Sylvia. 2001. Sinnstiftender Mathematikunterricht für Mädchen und Jungen. Schneider Verlag Hohengehren: Baltmannsweiler

Jungwirth, Helga. 1991. Die Dimension „Geschlecht" in den Interaktionen des Mathematikunterrichts. Journal für Mathematikdidaktik, 12 (2/3): 133-170

Jungwirth, Helga & Stadler, Helga. 2005. Gender-Sensibilisierung von Lehrkräften: Einstieg und organisierte Förderung durch die Fachdidaktik, in: *Die Forscher/innen von morgen*. Kongressbericht des 4. Internationalen Begabtenkongresses in Salzburg edited by Österreichisches Zentrum für Begabtenförderung und Begabtenforschung. Innsbruck/Wien/Bozen: 161-168

Kaiser-Meßmer, Gabriele. 1993. Results of an empirical study into gender differences in attitudes towards mathematics. Educational Studies in Mathematics, 2: 56-66

Kampshoff, Marita. 2007. Geschlechterdifferenz und Schulleistung. Deutsche und englische Studien im Vergleich. VS Verlag: Wiesbaden

Keitel-Kreidt, Christine. 2007. Geschlechtererziehung: Der blinde Fleck in der Lehrerinnenaus- und -fortbildung? In: Wissenschaftlerinnen-Rundbrief FU Berlin edited by Zentrale Frauenbeauftragte der FU Berlin. Nr. 3/2007: 4-10

Keller, Carmen. 1998. Geschlechterdifferenzen in der Mathematik: Prüfung von Erklärungsansätzen. Eine mehrebenenanalytische Untersuchung im Rahmen der ‚Third International Mathematics and Science Study'. Zürich
Köller, Olaf; Baumert, Jürgen & Neubrand, Johanna. 2000. Epistemologische Überzeugungen und Fachverständnis im Mathematik- und Physikunterricht, in: *Dritte Internationale Mathematik und Naturwissenschaftstudie – Mathematische und naturwissenschaftliche Bildung am Ende der Schullaufbahn,* edited by Jürgen Baumert et al. Leske & Budrich: Opladen: 229-269
Kriege, Jürgen. 1995. Die Rolle von Mädchen und Frauen in Schulbüchern – am Beispiel Mathematik, in: *Schule der Gleichberechtigung,* edited by Ministerium für Familie, Frauen, Weiterbildung und Kunst/Ministerium für Kultus und Sport Baden-Württemberg. Stuttgart: 169-173
Langfeldt, Bettina; Mehlmann, Sabine & Mischau, Anina. 2012. Genderkompetenz – (k)ein Thema in der universitären Lehramtsausbildung im Fach Mathematik, in: Klein, Uta & Heitzmann, Daniela (eds.). Diversity und Hochschule: Gestaltung von Vielfalt an Hochschulen. Juventa: Weinheim: 13-29
Lehmann, Rainer H. et al. 2000. QuaSUM Qualitätsuntersuchung an Schulen zum Unterricht in Mathematik. Ergebnisse einer repräsentativen Untersuchung im Land Brandenburg, Ministerium für Bildung, Jugend und Sport des Landes Brandenburg: Schulforschung in Brandenburg, Heft 1: Potsdam In: http://www.mbjs.brandenburg.de/sixcms/detail.php/5lbm1.c.152790.de [22.01.2013]
Lehmann, Rainer H. et al. 2001. LAU 9 Aspekte der Lernausgangslage und der Lernentwicklung – Klassenstufe 9. Ergebnisse einer längsschnittlichen Untersuchung in Hamburg. In: http://www2.bildungsserver.hamburg.de/contentblob/ 2815692/ data/pdf-schulleistungstest-lau-9.pdf [22.01.2013]
Leuders, Timo (eds.). 2003. Mathematik-Didaktik. Praxishandbuch für die Sekundarstufe I und II. Cornelsen Verlag Scriptor: Berlin
Mischau, Anina; Langfeldt, Bettina & Mehlmann, Sabine. 2009. Genderkompetenz als innovatives Element der Professionalisierung der LehrerInnenausbildung für das Fach Mathematik. Journal des Netzwerks Frauenforschung NRW, 25: 50-53
Mischau, Anina; Langfeldt, Bettina; Mehlmann, Sabine; Wöllmann, Torsten & Blunck, Andrea. 2010. Auf dem Weg zu genderkompetenten LehrerInnen im Unterrichtsfach Mathematik. Journal Netzwerk Frauen- und Geschlechterforschung NRW, 27: 29-39
Mischau, Anina. 2012. Genderkompetenz in der Lehramtsausbildung Mathematik: Möglichkeiten der Integration und Vermittlung. Wissenschaftlerinnen-Rundbrief, 1/2012: 19-22
Mullis, Ina V.S. et al. 2009. Gender differences in achievement. IEA's Third International Mathematics and Science Study. Boston
Nyssen, Elke. 2004. Gender in den Sekundarstufen, in: Handbuch Gender und Erziehungswissenschaft edited by Edith Glaser. Klinkhardt: Bad Heilbrunn: 389-409
OECD. 2004. Lernen für die Welt von morgen: Erste Ergebnisse von PISA 2003. Spektrum akademischer Verlag
OECD. 2009. Equally prepared for life? How 15-year-old boys and girls perform in school. In: http://www.pisa.oecd.org/document/51/0,3343,en_32252351_ 322361 91_ 42837811_1_1_1_1,00.html [22.01.2013]

Stanat, P.; Kunter, M. 2001. Geschlechterunterschiede in Basiskompetenzen, in: PISA 2000. Basiskompetenzen von Schülerinnen und Schülern im internationalen Vergleich edited by Jürgen Baumert. Leske & Budrich Verlag: Opladen: 249-269

Steinthorsdottir, Olof Bjorg & Sriraman, Bharath. 2008. Exploring gender factors related to PISA 2003 results in Iceland: a youth interview study. ZDM Mathematics Education, 40: 591-600

Stürzer, Monika et al. 2003. Geschlechterverhältnisse in der Schule. VS Verlag: Opladen

Tausendpfund, Markus. 2007. Höheres Interesse, schlechtere Leistung: Geschlechtsspezifische Leistungserwartung in der Mathematik und ihr Einfluss auf die Testleistung in der PISA-Studie 2003. Mannheim

Valtin, Renate. 1993. Koedukation macht Mädchen brav!? – Der heimliche Lehrplan der geschlechtsspezifischen Sozialisation., in: *MädchenStärken. Probleme der Koedukation in der Grundschule,* edited by Gertrud Pfister & Renate Valtin. Frankfurt a.M.: 8-37

vbw – Vereinigung der Bayerischen Wirtschaft e.V. (eds.). 2009. Geschlechterdifferenzen im Bildungssystem. Jahresgutachten 2009. VS Verlag: Wiesbaden

Ziegler, Albert; Kuhn, Cornelia & Heller, Kurt. 1998. Implizite Theorien von gymnasialen Mathematik und Physiklehrkräften zu geschlechtsspezifischer Begabung und Motivation. Psychologische Beiträge, 40 (3-4): 271-287

Zimmer, Karin et al. 2004. Kompetenzen von Jungen und Mädchen, in: *PISA 2003. Der Bildungsstand der Jugendlichen in Deutschland – Ergebnisse des zweiten internationalen Vergleichs,* edited by PISA-Konsortium Deutschland. Waxmann: Münster: Chapter 8: 211-223

ICT resources in STEM activities and ICT and STEM career prospects for young students: Gender-related issues[34]

Àgueda Gras-Velázquez, Alexa Joyce and Albert Gras-Martí

Abstract

In response to the question how can we encourage students and teachers to look forward to STEM classes, we consider whether ICT resources are a valid option. We also address the related problem of lack of female-students interest in ICT-related careers. We review briefly five transnational European projects managed by European Schoolnet that provide different visions and data, as well as suggestions, about how to answer the previous questions.

1   Gender differences in ICT-related professions and ICT use in STEM

Two related issues will be addressed in this paper: The attitude of female and male elementary and high school students towards using resources based on Information and Communication Technologies (ICT) in Teaching and Learning (T&L) activities related to their Science, Technology, Engineering and Math (STEM) classes, and the prospects of female students in pursuing an ICT career in future studies.

It is well known that the fraction of undergraduate students in STEM degrees is much smaller than what the present technically-based society demands, and that the relative percentage of women is well below the 50-50 ratio. This is particularly true in engineering degrees, specifically in ICT-based careers, which are in great demand in the academic, professional and industry markets of the contemporary knowledge-intensive based world (Lisbon 2000). In Europe only 22% of students aged 20 to 24 study STEM subjects (Durando, Wastiau, & Joyce 2009).

However, girls use and enjoy ICT tools. According to Gras-Velázquez, Joyce & Debry (2009), Becta (2008), for instance, shows that girls use technology more than boys for social networking and creative purposes, whilst

---

[34] This paper is based on the talk How to encourage students and teachers to look forward to Maths Science and Technology classes: five projects, five visions given at The Gender Perspective of Young People's Images of Science, Engineering and Technology (SET), 10-12 December 2009, University of Wuppertal, Germany by À. Gras-Velázquez

Forrester (2009) indicates that girls are predominantly "'joiners', 'spectators' and 'creators' online." However, the influence of ICT in T&L activities in STEM classrooms has not been much investigated. Still less well known is the use of ICT resources by teachers, in particular by female teachers.

The lack of interest in STEM subjects especially among women may be due to lack of motivation from teachers (Pollen 2009), stereotyped thinking of parents and teachers towards STEM careers especially concerning women (Gras-Velázquez, Joyce & Kirsch 2009), unattractive and 'overstuffed' curricula giving only a superficial approach to science and the scientific method, and limited views of the science profession as only consisting of white lab-coat jobs (Lipsett 2008).

This situation is not desirable, because it means a waste of talent for the industry and the academy, and a lack of opportunity for women to enter the job market in the field of ICT and STEM-related professions. This situation also impacts on future generations because girls and boys are influenced by parents, teachers, and society members, in general, as key role models. Valid questions are whether this lack of interest of female students towards ICT resources and applications starts at an early age, and whether use of ICT in STEM subjects may help both in increasing an appreciation for ICT careers and better understanding of difficult STEM subjects.

We shall contribute some data that may help in finding a partial answer to the research questions mentioned above. These data come from a small set of EU-funded projects developed in collaboration with many stakeholders in the Science Education and ICT field. European Schoolnet (EUN) participated in all the projects that we shall review here in co-partnership with groups of teachers.

EUN is a network of 31 Ministries of Education in Europe that has three main objectives:

- To support schools in bringing the best use of technology to learning.
- To improve and raise the quality of education in Europe.
- To promote the European dimension in schools and Education.

Since 2007, Ministries of Education have set STEM as one of the major thematic domains in which EUN should play a role. More than 15 different projects are currently running in the STEM areas, including awareness-raising campaigns in schools in specific subjects, policy studies, and validation projects. In particular, attention in EUN is placed in the promotion of use of ICT in education, in partnership with major IT firms. EUN supports the argument that a reversal of school science-teaching pedagogy from mainly deductive to inquiry-based, 'hands-on' and other innovative methods is necessary if we are to increase interest in science (Rocard 2007).

In the following section we shall first present the five EUN-participated projects that we shall use as the main source of data. In section 3 we shall present the general research questions that guide this study, and in section 4 we discuss some results.

2   Five related EUN projects: Outline

Out of the various projects that EUN has run recently, that have direct relevance for the topic of this paper, we concentrate on five: Inspire, Stella, e-Skills, the ICT Survey and Items. At least partially, these projects share the aim to find the reasons for lack of gender equity in ICT and STEM. For information about other projects that have some bearing on the topic, like Xplore Health, Xperimania, Aspect, inGenious, etc., see the webpage of Scientix[35], http://www.scientix.eu, a site where the Science Education community meets and exchanges information. Brief descriptions of the five projects follow.

Three projects look into ICT use in STEM environments: **Inspire** (Innovative Science Pedagogy in Research and Education, http://inspire.eun.org) challenged the lack of interest of students to start scientific studies and, more widely, to extend the supply of scientific specialists and develop a scientific culture in European countries. Inspire gathered data to a) *observe* the impact of new teaching methods on pupils and on their motivation, b) to *analyse* the pre-requisites to be defined for enabling teachers to integrate new techniques in their pedagogy, and c) to *identify* the critical success factors to be mastered at teacher and school level for the generalization of such practices. The Inspire methodology was to collect a good number of learning resources (LR) in digital form and to make them available to 62 schools in 5 countries. Minimal language translation of LR was provided as well as minimum instructions and suggestions for classroom use. Online questionnaires were used to collect data on LR use and recommendations.

Secondly, on the dissemination front we shall use data from the **Stella** project (Science Teaching in a Lifelong Learning Approach, http://stella-science.eu), a multilingual web portal for those involved in science education

---

35   Scientix – the community for science education in Europe – was created to facilitate regular dissemination and sharing of know-how and best practices in science education across the European Union. Scientix is open for teachers, researchers, policy makers, parents and anyone interested in science education. The Scientix portal contains basic information on public funded projects, forums, chats, news and newsletters and most importantly, a repository of materials including lesson guides, animations, simulations, reports, videos, etc which teachers can use in their classes form different projects. The key difference of the Scientix repository is its translation on demand service. Teachers may request the free non-machine translation of the materials to any of the 23 official languages of the Commission. This translation is carried out by official translators and is completely financed by the European Commission. Scientix is financed under the European Union's Seventh Framework Programme for Research and Development and is managed by European Schoolnet (EUN) on behalf of the European Commission. More information: http://scientix.eu

to communicate experiences, cooperate, exchange ideas and thoughts on teaching methods and approaches. In particular, with relation to the topic of this paper, the project aimed to promote passion and interest in science subjects especially amongst young females.

And thirdly, the **Items** project (Improving Teacher Education in Math and Science, http://itemspro.net) is a Lifelong Learning Programme that developed frameworks for improving the competencies of S&M teachers and, consequently, to increase students attainment and interest in these areas. It provided modules for teachers, teachers training courses, and schools.

The other two projects that we consider are more industry oriented. The European **e-Skills** Career Portal project (http://eskills.eun.org) was supported by the e-Skills Industry Leadership Board that includes the Career Portal Committee and ICT firms. E-Skills launched campaigns to raise awareness of the growing demand of highly skilled ICT practitioners and users within the industry. An example of a research and dissemination project is a study of media images of Science, Engineering and Technology (SET) influencing young people's perception of these fields.

And finally the **ICT Survey** examined teenage girls and boys' attitudes to ICT and ICT careers in secondary schools in five countries in Europe, in order to verify whether there are differences in perception and/or aptitude between the genders and to understand what might be putting girls off further studies and careers in ICT by: A) looking at the impact of role models on study and career choices, b) assessing to what extent negative stereotypes affect girls' career choices in relation to ICT, and c) develop recommendations on the basis of the research.

Table 1: Example of countries, students and teachers participating in the ICT survey.

	Students	Teachers
NL	119	35
UK	98	31
IT	110	32
PL	94	30
FR	121	38
total	542	166
average	108.4	33.2

Source: Gras-Velázquez, Joyce & Debry (2009)

Typically, in EUN-managed projects many countries and hundreds of participants are involved, Table 1. This ensures a wide variety of cultural differences.

3 Research questions

The basic research questions that we address in this paper are the following:

- How can we encourage female students and female teachers to look forward to Mathematics-Science-Technology classes?
- Can ICT resources help in that aim?
- Why are girls less attracted to ICT-related careers?

We shall relate to the results of the projects described in section 2 which:

- Examine teenage girls and boys' attitudes to ICT and ICT careers in secondary schools.
- Analyze differences in perception and/or aptitude between the genders.
- Look into what puts girls off further studies and careers in ICT.
- Consider students opinions about LR in STEM: Is it true that girls also enjoy the use of LR in their classes?

- Study the transferability of LR in different school cultures.
- Develop recommendations on the basis of the research.

4 Results and discussion

In the following we shall briefly present the more relevant conclusions of the five projects described in section 2, regarding the research questions mentioned above. Detailed arguments and data will not be given for lack of space and the interested reader may refer to the original reference.

Girls still held back by stereotyped thinking, but attitudes may be evolving: Most girls drop out of ICT studies after secondary education. This can be attributed partly to lack of support from role models, persistent stereotyped views that the sector is better suited to men, a lack of understanding about what ICT jobs entail, and in some cases, how easy or difficult they find the subject. However, a key finding is that girls generally like and enjoy ICT studies and are competent users of computers and computer operating systems. Girls are roughly equal to boys in aptitude in ICT at secondary level and most girls enjoy studying ICT. However this enjoyment does not often transmit into careers.

What puts women off ICT? Girls and boys show differences in how they perceive computer science studies and careers. Girls more often associate the concept of ICT with hardware, algorithms and programming; whereas boys are more likely to see ICT as socially-oriented.

Female students like/dislike of ICT at school: In some countries girls expressed dislike for ICT as a subject, but girls see positively cooperative school projects in other countries. Despite having equally good – or better – grades as male counterparts in STEM subjects, girls are often actively discouraged by families, teachers, and career advisors from pursuing further studies or careers in the field.

We see in Figure , from the ICT survey, the noticeable differences among various countries (x-axis) with regards to liking ICT in school and planning an ICT-related career.

Figure 1: Comparison on ICT attitudes for various countries. Shaded columns indicate women who believe ICT sector to be inherently better suited to men, and blue columns (on top of the former shadowed columns) is the percentage of female students interested in ICT. Red (top) line is % of women who find ICT easier than other subjects; black (lower in IT) line is % of women who intend to continue studying ICT careers or professions. Green (middle in IT) line is % of women interested in Internet networking careers.

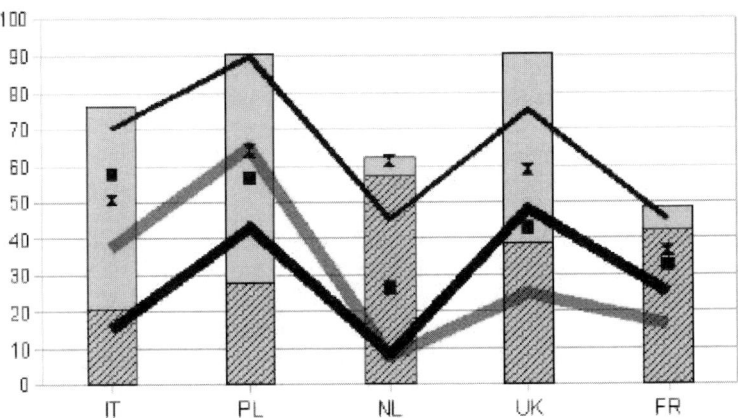

Source: Gras-Velázquez, Joyce & Debry (2009)

Girls tend to study foreign languages and are interested in travel and see a mismatch between perceived job attributes and what they want from their 'dream job': Girls in all countries expressed the desire to help others, to travel and to have a high degree of autonomy and independence in their future careers. When asked about ICT, they did not expect to find roles that would fulfill these aspirations. Measures should be taken to reverse this perception.

Teachers and parents consider that ICT and ICT careers are better suited to men: In all countries, teachers and parents typically hold stereotyped views of the sector. It is of key importance to give parents and teachers a more balanced perspective via better support for school staff such as teachers and career advisors, to give them a clearer idea of IT career options, e.g., awareness-raising materials and campaigns, training sessions with IT companies, and 'open days' for local schools to visit IT facilities.

Figure 2: Impact of the LR on a) the Lithuanian (left) and b) Austrian (right) students, filtered by gender.

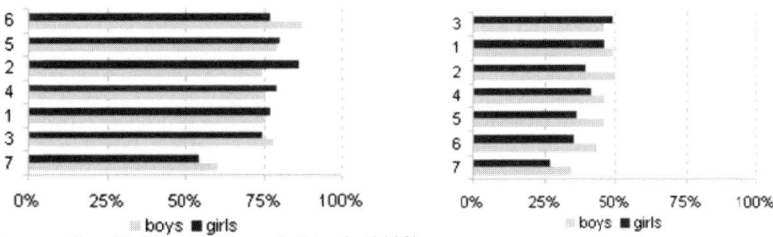

Source: Gras-Velázquez, Joyce & Kirsch (2009)

1 = LR made it easier to integrate/remember STEM lessons; 2 = LR made it easier to understand and learn STEM; 3 = LR made it easier to study by myself; 4 = LR made it easier to understand the use of ICT; 5 = LR stimulated interest and motivation for STEM; 6 = LR made it easier to link STEM to my everyday life; 7 = Helped me choose my profession.

Assessment of the impact of the use of LR in class: Thousands of students participated in a survey to assess their appreciation of STEM and corroborated studies on male and female students' perceptions of STEM and ICT (PISA 2000-2006; Gras-Velázquez et al. 2009b) in that the impact of STEM subjects is generally more prominent with boys than with girls. The biggest impact STEM lessons have is on their choice of careers, with 74% of the boys seeing these subjects as weighing on their selection of future career path, compared to only 60% of the girls. Additionally, 73% of the boys in the survey were interested in and motivated for STEM compared to only 53% of the girls. Other interesting results were that 66% of the boys and 58% of the girls found it easy to study STEM with LR by themselves at their own pace. We compare in Figure results from two countries (Lithuania Figure a, and Austria Figure b). It is interesting to note that the gender differences seen in the other countries are not as marked in the results from Lithuania. Lithuanian female students found the resources increased their interest in and understanding of STEM and even in ICT more than their male counterparts. For this country, 79% of the girls were interested in STEM, compared to 67% of the male students. The decrease in the impact with age is present in all countries. The drop-off is more acute among girls, perhaps due to increased pressure of gender stereotypes.

The Inspire results from teachers were good in terms of:

- Highest impact in autonomous learning of pupils.
- 75% of LR stimulated own interest and motivation for teaching STEM.
- 70% said that the LR increased the pupils' understanding and use of ICT in general.
- 2/3 noticed that the LR stimulated pupils' interest and motivation for learning STEM.
- Facilitated differentiated teaching of sciences in the classroom.

In terms of the present paper, one should remark that the Inspire project proved to have a greater impact on boys than on girls, and the impact of LR probably decreases with age, especially among female students, as shown by the ICT survey. The reasons for both these behaviors were not quite clear and merit further investigations.

Figure 3: Influence of female role models on female students.

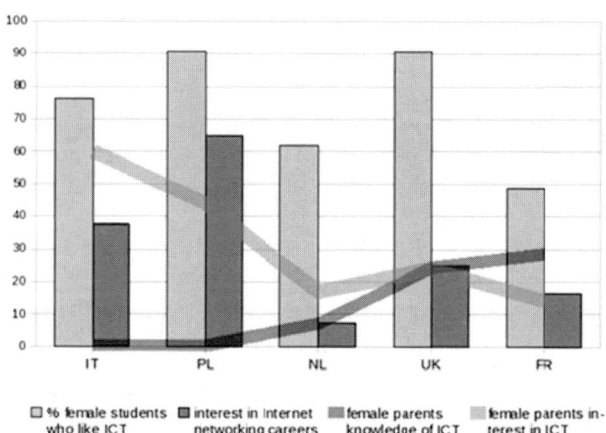

Source: Gras-Velázquez, Joyce & Debry (2009)

Girls lack positive ICT- oriented role models: In all countries, female students are more influenced by role models than their male counterparts, and in particular, turn to female role models when making career and study choices, *Figure* . The importance (and lack of) role models for women pursuing careers in STEM and ICT fields is a recurrent finding in all projects. This point is specifically addressed in some projects, like Stella, that provides a section with interviews especially with (women) successful experts in ICT jobs and discussion boards for women to promote discussion and share information about ICT. Energizing women in ICT to blog and video blog would allow for the socialization of role models: A scarce resource in this field.

Although liking a subject does not necessarily mean pursuing a career in that subject, Figure shows that when boys and girls are questioned about which subjects they planned to continue studying, the fraction of girls mentioning Physics and ICT was 50% than that of boys. The behavior in the rest of the subjects is also as expected: Women prefer languages and literature, and also design-technology.

Figure 4: Future career choice according to gender. Lines: Students' relative interest in the different subjects (calibrated to match "columns" scale).

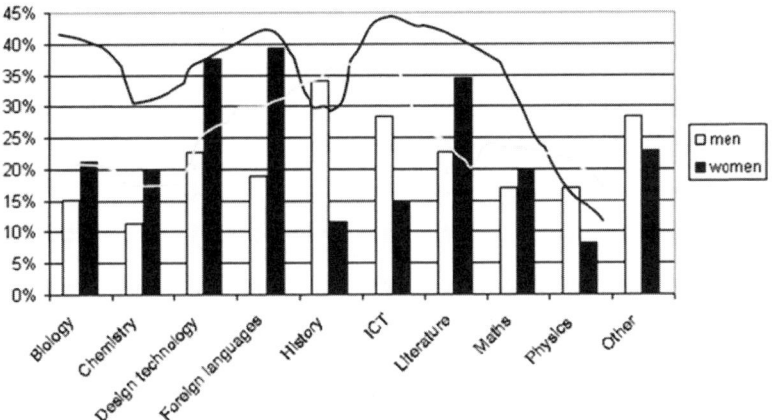

Source: Gras-Velázquez, Joyce & Debry (2009)

A comparative analysis of the country findings yielded the following points: Polish female students have the most positive view towards ICT, ICT jobs and Internet networking careers, followed by Italy and the UK. Dutch female students have the most negative view and see ICT networking careers as better suited to men. In general, 50% fewer female students are interested in studying ICT in the future compared to the percentage that report liking ICT at school. Except in the Netherlands, over 50% of students are influenced by role models (parents, celebrities and teachers). In particular, male students look to male role models, and girls to female role models. In general, the research found positive attitude towards ICT from female parents, especially in Poland, Italy and France − although this was much less the case in the Netherlands.

On a final note we may mention briefly an investigation of the perception and behavior of women in online courses (Gras-Martí et al. 2009). Although this research is not part of any of the five EUN-managed projects discussed so far, it bears much importance with regards to ICT environments and the advantages they offer to female students. Online learning activities (like forums, for instance) have the advantage that women are not in the presence of male-dominated classrooms (in STM studies). It turns out that women operate easily in online environments, manage their time better and participate more thoroughly than men, also contributing with deeper and better supported arguments.

5 Conclusions

From the projects revisited in this paper one may draw a number of conclusions and proposals for improving ICT use among female students and ICT and STEM career prospects among girls. The pattern of under-representation of women in ICT and STEM is set to continue if more is not done to educate, support and encourage girls and their role models. We conclude that public-private collaboration could play a role in changing perceptions about industry, by giving access to more realistic and authentic information about STEM and STEM careers. In particular, closer cooperation is needed among education agencies and Ministries, together with industry, to ensure accurate information about ICT and STEM careers is available to teachers, pupils and their parents. Numerous initiatives have been launched, but the mainstreaming of such initiatives is required to have a systemic impact. Furthermore, the evaluations of school-industry initiatives like those that will be carried out by the inGenious[36] project will furthermore help design more effective initiatives in the future.

The importance of women role models for increasing ICT and STEM attraction for girls is stressed in various projects. Furthermore, a follow-up on gender-based impact is needed in projects. These would address basic questions like why girls like LR less than boys, would look at STEM classrooms, and question what are the differences are in LR use among male and female teachers. Projects like Scientix, where its portal has a repository of LR continuously enhanced with materials from all European STEM public funded

---

36 inGenious, the platform of the European Coordinating Body in Science, Technology Engineering and Mathematics (STEM) Education. It is a joint initiative launched by European Schoolnet and the European Roundtable of Industrialists (ERT) aiming to reinforce young European's interest in science education and careers and thus address the future skills gap. Through a strategic partnership between major industries and Ministries of Education, inGenious has the objective of increasing the links between science education and careers, by involving up to 1,000 classrooms throughout Europe. All the actions undertaken in ingenious will ensure that education / industry cooperation initiatives improve the image of STEM careers among young people and encourage them to think about the wide range of interesting opportunities that STEM can bring to their lives in the future. inGenious supported by the European Union's Framework Programme for Research and Development (FP7). More information: http://www.ingenious-science.eu/

projects, could provide additional information at least as far as the use of female teachers of these resources.

Finally, we must mention that the new trend of using ICT tools in the study of humanities (like virtual reality in history classes, for instance) may have an indirect effect on girls, who are typically the majority of the student population in those areas. But, overall, the main message is that in order to address transversal issues like women's relation to ICT and STEM there are no partial solutions and a combination of approaches from various fronts is needed, involving all societal stakeholders: Schools, teachers, industry and European policy managers.

References

Becta. 2008. How do boys and girls differ in their use of ICT? Research report. Retrieved 1 June 2009 from http://partners.becta.org.uk/upload-dir/downloads/page_documents/research/gender_ict_briefing.pdf

Durando, M.; Wastiau, P. & Joyce, A. 2009. Women in IT: The European situation and the role of publicprivate partnerships in promoting greater participation of young women in technology Special Insight Report (Online) Available at http://resources.eun.org/insight/Science%20girls5.pdf [Accessed: 20 July 2010]

Forrester. 2009. Groundswell profile tool. Retrieved 1 June 2009 from http://www.forrester.com/Groundswell/profile_tool.html

Gras-Martí, A.; Mora Torres, E.; Gras-Velázquez, À. & López, M. L. 2009. Estudi de cas sobre prespectives de gènere en els debats virtuals. Feminismo/s, Núm. 14: Género y nuevas tecnologías. CED, Universitat d'Alacant: 71-86.

Gras-Velázquez, À.; Joyce, A. & Debry, M. 2009. Women and ICT, Why are girls still not attracted to ICT studies and careers? Cisco White paper (Online) Available at http://eun.org/whitepaper [Accessed: 7 November 2009]

Gras-Velázquez, À.; Joyce, A. & Kirsch, M. 2009. Inspire: Challenging the lack of interest in MST among students using Learning Resources. Insight report (Online). Available at http://inspire.eun.org/index.php/Publications

Lipsett, A. 2008. Teens do not see science as route to good career. (Online) Available at: http://www.guardian.co.uk/education/2008/nov/07/science-careers-hamilton [Accessed: July 2009] Lisbon (2000). Lisbon European Council 23 and 24 March 2000, Presidency Conclusions http://www.europarl.europa.eu/summits/lis1_en.htm

Lisbon. 2000. European Council 23 and 24 March 2000, Presidency Conclusions.

PISA. 2000-2006. Results of the international studies PISA 2000-2006, reports. (Online) Available at: http://www.pisa.oecd.org [Accessed: 15 October]

Pollen. 2009. Guide for the Seed City Trainer (Online) Available at: http://www.polleneuropa.net/telecharger.php?rep=A4GBpvswB5RQFIr51PVPFQ%3D%3D&nom=3OLu8tmjtbP%2B9vqapVDTWO5YQj%2F0YX9c [Accessed July 2009]

Rocard et al. 2007. High Level Group on Science Education, Directorate General for Research, Science, Economy and Science, European Commission, Science Education Now: A Renewed Pedagogy for the Future of Europe: http://ec.europa.eu/

research/science-society/document_library/pdf_06/report-rocardon-science-education_en.pdf

Part 3 – Methodology and evaluation of good practice

The third part is focusing on implementation and reflection of initiatives for attracting young people for SET. Papers are presenting initiatives as well as evaluations even though their success and impact often is not directly measurable. They critically discuss under this topic the starting points and effects of so called inclusion measures aiming on changing the image of SET. Questions for discussion are: How are measures for motivating and attracting more young people for these subjects working? Which approaches do they use and what is their concept behind? And finally, what methodological questions arise for evaluating such inclusion initiatives?

Are young people lazy, blind, or misguided?
A fresh look at unpleasant facts

Frank Stefan Becker

The quest for engineers during the past boom and predictions of future shortages have focused attention on the low enrolment figures in science and technology (S&T) subjects. Normally, it is assumed that young people shy away from "tough majors" or make irrational choices, based on an absence of information. While not denying the fundamental necessity that a higher proportion of the population should have a background in science and technology, I want to pursue a different approach. Only by identifying potentially valid reasons for the lack of interest in S&T will it be possible to change not just some "misguided" perceptions among the younger generation, but to categorize the facts and make targeted recommendations for necessary changes. Therefore, this article will discuss the importance of image and status, the influence of society and peer groups, as well as financial rewards and career aspects. It will be shown that the universally observable trend away from S&T is not due to a dislike of technology on the part of the younger generation, but is caused by the fact that careers in this field don't seem attractive enough, especially in comparison to alternatives available in developed countries. Some recommendations to improve this situation are offered.

1 Introduction

An actual or impending shortage of engineers was the cause of much concern in the boom years before the financial crisis (e.g. Johnson 2006; Winckler 2006; Goossens 2007). The phenomenon is a widespread one (EU 2008), especially in the industrialized countries (Sjøberg 2010) and has become a hot topic again as the economy recovers. The reason most often cited is the unwillingness of young people to pursue careers in engineering (Figure 1). In Germany, for example, the share of students enrolled in electrical engineering has decreased from 5.8 percent in 1990 to 3.2 percent in 2009 (Becker 2011b). Other factors include the high dropout rates among engineering students – rates typically ranging from 25 to 50 percent at German universities, for example – and the persistently low number of women enrolled in engineering.

The situation, which is unsatisfactory for society as well as for industry and for students, has been the subject of numerous, recently summarized studies, which, however, "*still largely focus on symptoms*" (Prieto 2009). At the same time, it has triggered a plethora of initiatives aimed at fostering

interest in science and technology – over 900 in Germany alone (MoMoTech 2009) and many more in other countries (Tengelin 2009). As the desired improvements have largely failed to materialize – that is, young people have persistently not reacted as hoped – there appear to be mainly two possible explanations:
- Either young people are too dumb to understand the advantages of an engineering career, *or*
- they are too clever to overlook its disadvantages.

Figure 1: Willingness of young people to study engineering in selected European countries

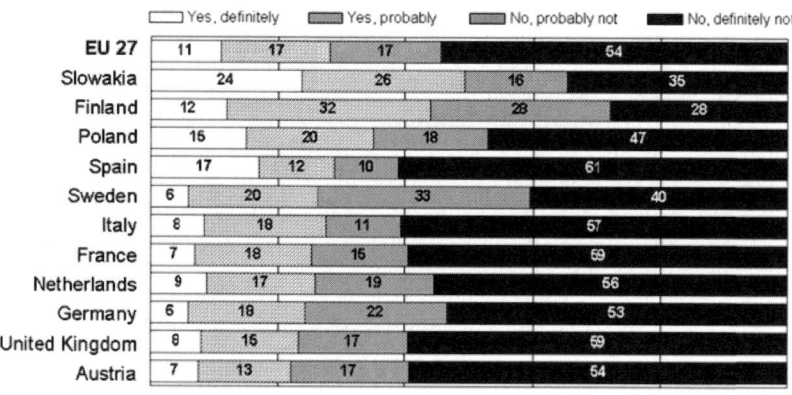

Source: EU 2008

Discussion to date has focused – if only implicitly – on the first assumption, which in turn has triggered even more well-intentioned efforts at persuasion. However, as the aim cannot be to talk unwilling or ungifted youngsters into engineering, I will try to explore the second alternative – even at the risk of debunking some common myths and presenting unpleasant truths. If change is really desired, it is on us, the "adult" decision makers in society, to dispense with wishful thinking and try to respect the reasons the target group may have for disregarding our unanimous advice. Only then will it be possible to identify the real obstacles and to categorize them according to "cannot be changed," "can only be improved in a wider societal context" or "has to be changed by … schools, universities, or companies etc." To this end, this article will analyze a host of studies, including the results of a recent large-scale survey undertaken in Germany (NaBaTech 2009) as well as anecdotal, mostly German evidence.

2 The changing role of engineers and technology

The second half of the 19th century was the age of engineers, with pioneering figures like Edison and Siemens. In an environment still largely technology-free, the challenge then was to develop working innovations and not – as it is today – to prevail in an intensively competitive market, where a wide array of non-technical factors determine success. Engineering brought progress visible to all: Railways, electric lighting and motors in factories, automobiles, steamships, radio communications, X-rays, telephone links, airplanes and a host of other inventions.

As a result, engineers and natural scientists could be the heroes of popular, technology-oriented novels such as those of Jules Verne in France and Hans Dominik in Germany, which had a readership of millions. Despite the misuse to which technology was put in World War II, the faith in progress that inspired those novels remained widespread until the late 1960s. Since then, however, a reaction against science and technology (S&T) has set in, fueled by the increasing opaqueness of modern technologies and the rise of the environmentalist movement. Visions were lost – that is, the belief waned *"that the future should be better than the past"* and *"that everything can and should be improved"* (Winckler 2006). This has significantly impaired the reputation of engineers, as the clear link between new technologies and better living standards has become blurred – a connection still evident in today's industrializing nations. Here, the percentage of young people wanting to pursue careers in engineering is much higher, as illustrated in Figure 2. This difference is especially pronounced for girls, a finding which will be discussed in section 7.

In the developed countries, omnipresent technology has become a normal element of everyday life – something merely used ("switched on") and taken for granted. This high-tech saturation is paralleled by the "invisibility of technology" – that is, the reliance of performance on hidden features like microelectronics and software. Most highly qualified work – be it text production, administration tasks, graphics design, factory simulation or the development of new turbines – now involves primarily the computer screen. As a rule, direct, hands-on technology experience is nearly impossible in the everyday environment – thus eliminating a strong incentive for pursuing it.

Although the importance of S&T for the foundation of society as well as for everyday life has continued to grow, the gap between technology "nerds" and technology users has widened. Consequently, starting in the 1980s, an increasing number of more or less technically trained users – who were able to convert the new opportunities created by microelectronics, software and later the Internet into successful business models – rose to stardom. The signal was not lost on the younger generation, whose interest in science, technology, information technology and mathematics (STIM) developed – as

Figure 3 shows for Germany – largely parallel to the importance of research-intensive industries.

Figure 2: Willingness of young women and young men to become engineers in different countries. Countries with the highest Human Development Index are lowest on the vertical axis.

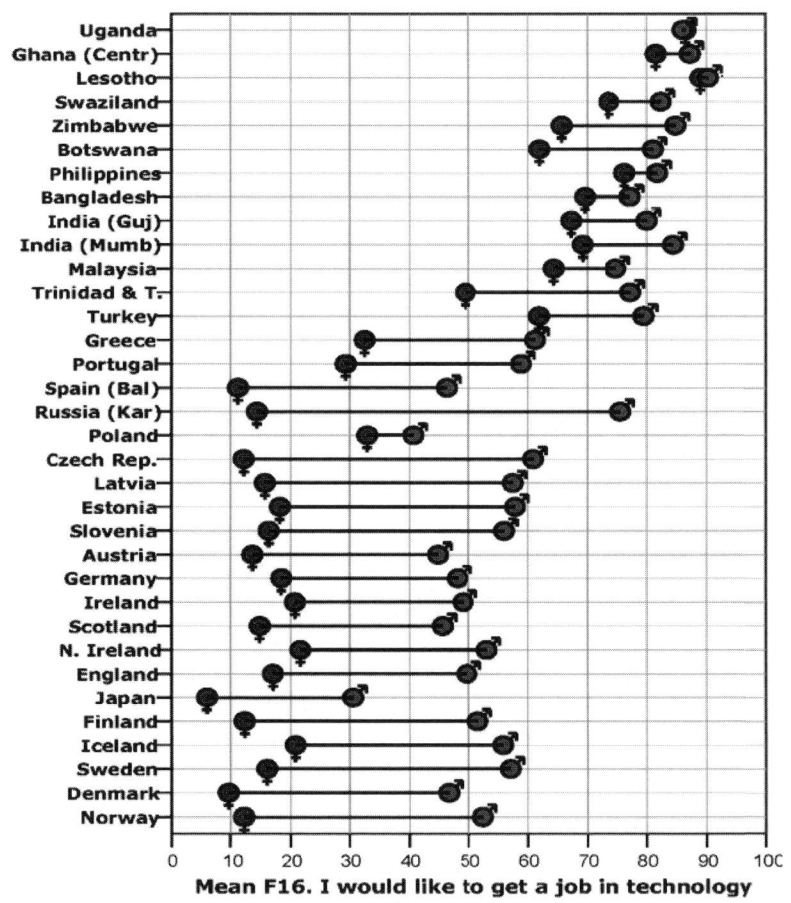

Source: Sjøberg 2010

However, the slight rebound of the share of German industry as a percentage of gross domestic product (GDP), noticeable since the mid-1990s (Meyer 2008), is a specific phenomenon not to be observed in most developed economies. For this reason, the relatively small percentage of young people in practically all OECD countries opting for engineering and science can be interpreted as a rational reaction to the trend toward a service economy in which the skills increasingly demanded are not taught in conventional engineering programs.

Figure 3: The enrolment in science, technology, IT and mathematics (STIM) majors parallels the importance of the R&D- intensive industry in Germany.

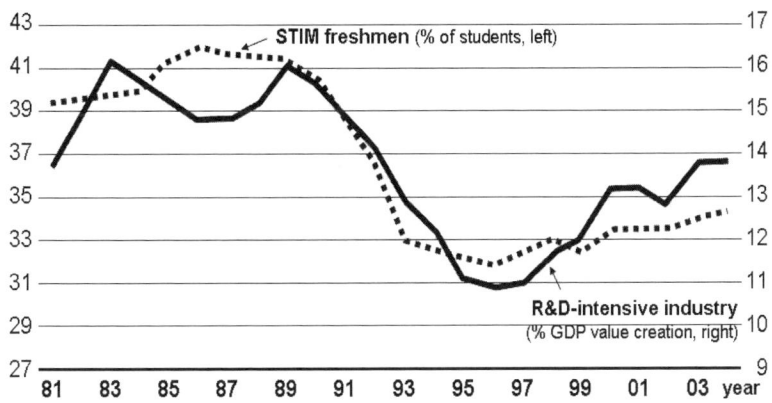

Source:(Meyer 2008

3 Fascinating technology – but off-putting university programs

The debate about the content of engineering courses is age-old (Becker 2009), and curricular reforms advocated by the business community have been as frequent as they have been unavailing (e.g. Becker 2006; Becker 2009; Becker 2010; Becker 2011a; Becker 2011b; Becker 2011c; Becker 2012a; Becker 2012b). The hopes for a more practice-oriented and career-enabling undergraduate curriculum that had been raised by the Bologna-Process in Germany have been only partially realized. Traditional university representatives often express doubts about the acceptance of undergraduates in the labor market. However, their real concern – dictated understandably by their own interests – is that they might lose their best students to industry before they can contribute to university research. Professor Gerlach (2007) from TU Dresden described the danger:

"Look at what goes on in the United States. Headhunters use money to entice the top students with bachelor's degrees from U.S. colleges into industry. Many students who should have gone on to graduate school, completed PhDs and gone into research don't because they've been lured away by the prospects of high salaries."

And Professor Henning (2007), dean of the faculty of Mechanical Engineering of the RWTH Aachen stated: "We don't want students to leave the university with only a bachelor's degree. But I'm afraid that, when engineers are in short supply, industry will entice them away." In line with these statements, 83 percent of German university professors did not consider the bachelor to be a standalone degree but "rather an intermediate step on the way to a master (Minks 2007). The TU9, a pressure-group of nine traditional German technical universities (among them Aachen and Dresden), have publicly claimed that only a master's degree should be equivalent to the former German "Diplomingenieur" degree – in spite of the fact that about 60 percent of all German engineers are graduates of Fachhochschulen (Universities of Applied Sciences), where the previous standard degree, the "Diplomingenieur (FH)", has been transformed into a bachelor's. The TU9 even goes so far as to demand that "the master's degree be the standard qualification for careers in academics **as well as business**." (TU9 2009). Such claims, which turn reality upside-down, have nevertheless done much to discredit the bachelor's degree in the eyes of an ill-informed public, of the students as well as potential employers. The negative effects of this position are twofold:

- The high threshold (minimum 5 years) for achieving this "real" master's degree puts off most young people who are interested in science and technology, but do not want to commit themselves for such a long time exclusively to engineering.
- As the bachelor's curriculum is not optimized to be a practical "degree relevant for the labor market", but mainly the theoretical foundation for the master's, it fails to fulfill the expectations of students who have chosen engineering because they are attracted by technological applications (NaBaTech 2009).

As a consequence, the old, well-known problems of the traditional diploma that have produced high dropout rates in the early study phase are bound to continue.

"For historical reasons, engineering students at German universities spend their first four semesters learning mostly theory. Practical applications are taught only later. As a result, the first years are more frustrating than enjoyable. Pedagogically considered, this approach violates all the basic principles of education, since it is questions of practical application that arouse an interest in theory and not the other way around."(Pritschow 2004).

Added to this is the elite attitude of many professors who believe that the study of engineering is and must be something particularly difficult! The president of the Technical University of Munich stated in an interview (Herrmann 2010):

"Those who do not make it through our system fail because of the demanding mathematics and the demanding sciences, because they do not acquire the level of theoretical knowledge that we and the international competition demand."

Instead of making mathematics the focus of good teaching, however, it is often used to get rid of students deemed unqualified to become a top-notch researcher, a trend well known in the field (Winkelman 2009). *"There must be something that can serve as a quantifiable criterion for failure"* was the explanation given by Herrmann (2011) when asked about the role of mathematics regarding the high drop-out rates of engineering students at the university.

The same spirit has led the rector of another leading German technical university to attribute the declining interest in engineering to young people's *"unwillingness to suffer"* – but not, of course, to unattractive teaching! It only remains to explain why these "suffering-averse" young people then go on to pursue degrees in other fields such as economics, law and medicine, which hardly involve a walk in the park...

To be fair, it should be noted that the government does not support engineering teaching to the extent required. For example, between 1995 and 2005, 10.6 percent of such "expensive" professorships were eliminated at German universities of applied sciences and even more, 13.3 percent, at German universities (IW 2008).

But high dropout rates – due also to financial difficulties and the temptations of the labor market (VDMA 2009) – are by no means the only negative consequences of an intense, theory-laden curriculum that leaves those students who cannot complete the minimum ten semesters of work required for a master's degree in their own eyes without a sufficient qualification. The deterrent created by this high threshold is so serious that it has helped make engineering the preserve of specialists with narrow interests, while industry cries out for problem-solvers with wide horizons (Becker 2011c; Becker 2012a; Becker 2012b). At the same time, it scares away young women, a topic dealt with in section 7. Studies show that precisely those young people who, because of their broader natural abilities, can pick and choose which career path they want to follow, tend to turn their backs on engineering (Minks 2005), especially as the profession has lost its reputation of providing secure jobs.

4 Reliability of predictions: Retrain engineers as plumbers!!

"Like every market, the market for engineers is characterized by imbalances between supply and demand. Both empirical observations and theoretical considerations indicate that, in the normal case, this relationship consists in a systematic mismatch – that is, a fit that is inexact both in terms of numbers and in terms of qualifications. An overview of labor market cycles over the past 23 years shows us, as empirical university and job researchers,

- that projections and prognoses of the future demand for engineers have proven to be imprecise and mistaken and

- that, following their announcement, these forecasts have caused participants to "overreact," which, in turn, has often led to a disparity between the supply of and the demand for engineers – a frequently noted divergence that intensifies from cycle to cycle." (Winkler 1996).

This sobering assessment was made when – in the wake of an economic crisis and the collapse of engineering-based industries in the former East Germany – whole age-groups of German university graduates found themselves without employment. They had relied on rosy predictions of a promising future:

"In the middle of the 1990s, the number of available graduates in the area of electrical engineering will no longer be sufficient to meet the needs of industry for young electrical engineers. Educational policymakers and industry associations based this conclusion on the estimated student numbers, on the one hand, and on the demand for electrical engineers expected in the mid 1990s, on the other" (ZVEI 1987).

Much more dramatic was the explosion in the numbers of unemployed engineers over the age of 40, which, in the case of electrical engineers, for example, exceeded 300 percent. Of course the economic slump also impacted other professions and no one could have predicted the sudden reunification of Germany. (That such disruptive events must always be taken into account, however, is shown by the terrorist attacks of 2001 and the financial crisis of 2008). But for security-oriented engineers and IT specialists, who frequently came from upwardly mobile families (Hartmann 2009; 4Ing 2009), this was a new and particularly bitter experience (Becker 2009) that still lingers in the "collective memory," as will be discussed in section 8. In the nineties, young people got the message and – as Figure 4 shows – turned their backs on engineering studies in droves.

Their market-oriented reaction was in line with the statements of political leaders and industry associations. A "Forum for Innovation" organized in 1997 came to the resigned conclusion that *"no positive job developments in the conventional fields of engineering and the sciences are to be expected in the foreseeable future."* At the Forum, leading specialists discussed *"how the transition from engineers and natural scientists to alternative fields can be supported."* In this connection, the *"engineers-as-craftsman-model"* was touted as particularly promising (Staudt 1998).

But within only two years, the world had changed once more, reports on the lack of IT experts filled the media, and university graduates were now in short supply (Figure 4).

Mismatches between the number of qualified university graduates and the number of job openings will be impossible to avoid in the future since, as Figure 5 shows, companies' hiring patterns are generally linked to market cycles.

Considering this connection, claims that the EU may lack as many as 700,000 researchers are to be treated cautiously. Such claims arise from the politically motivated, but far from realized goal of increasing R&D expenditures in the future to three percent of total EU GDP from their current 1.9 percent.

Figure 4: Development of freshmen and graduates in electrical engineering in Germany. Enrolment is strongly influenced by economic perspectives.

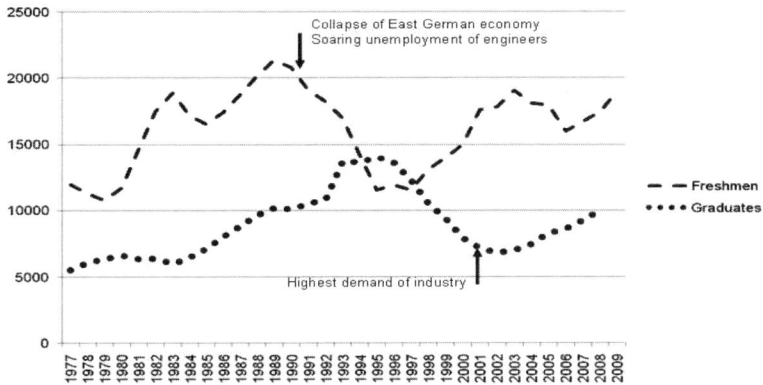

Source: ZVEI

Figure 5: Hiring of new employees with a technical university degree at Siemens AG and corresponding changes in the Siemens-relevant German electrical market.

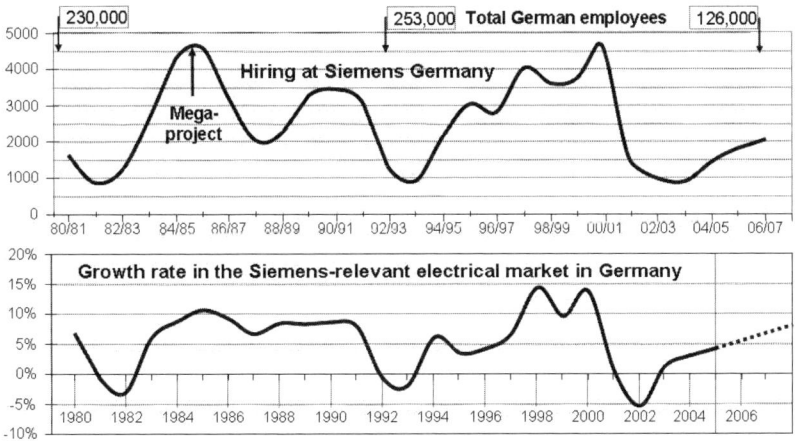

Source: Becker 2008

181

As a general rule, however, university graduates with degrees in science- and technology-related subjects ought to have above-average job entry prospects even in difficult times. The prerequisite, though, is that students are not fixated on a single specialist career path but regard their university education primarily as a training in the ability to think. This attitude is much more common in English-speaking countries, where the connection between college major and career path is looser (for example, Abbot 2003). The rigorous focus on specialist training in Germany's technically excellent engineering schools is both one of their strengths and one of their fundamental problems – especially if graduates want to move up in the corporate hierarchy.

5   Engineering – the fast career track to the top?

Complaints about the one-sidedness of engineering studies have a long tradition (Becker 2009). In this respect, the transfer of a large number of German science and technology departments to remote and secluded "science ghettos" in the 1960s (for example, Marburg, Tübingen, Munich-Garching) was detrimental to the goal of fostering the wider perspective required in a modern service economy.

In Germany today, the system for educating engineers is quite successful in promoting theoretical expertise but less so in meeting the other job requirements, as surveys of young engineers regularly reveal. The information presented in Figure 6 is representative of many analyses. The picture is a bit less dramatic if the softer "important/rather well" categories are included too, but the general impression remains (Becker 2009).

The ease with which engineering graduates have been able to find jobs in recent years seems to belie these findings or render the discrepancies irrelevant, but we will revert to this topic later. In fact, the good job offers for entry-level positions, especially in 2006-2008, are often cited as an argument for the profession's attractiveness, and salaries for engineers can be 25 percent higher than for graduates in the humanities (iw 2008; Fabian 2009). As this situation has still not, however, triggered a stampede into engineering, the situation calls for closer examination.

In a market economy, a shortage usually manifests itself by increasing "prices," that is, salaries – an effect that could motivate young people to choose the profession. But can this really be observed in the labor market? Unfortunately, the available data do not support this conclusion.

In Germany, the development of entry-level salaries for engineers in the industries, shown in Figure 7, if adjusted for the cumulated inflation of 15.5 percent (7/2002-7/2011), indicates, indeed, a distinct drop in real-term buying-power in most sectors (VDI 2011). These findings are confirmed by a study analyzing the general salary development of engineers between 1990 and 2008 (VDI 2010). Other professions may have suffered declines as well – but here we are talking about a highly qualified group allegedly in urgent demand! It doesn't take much to see that in the eyes of our target group, which is also able to calculate, the much-debated shortage of engineers has failed to produce the logical positive consequences.

Figure 6: Disparity between knowledge taught at universities and know-how required in the workplace

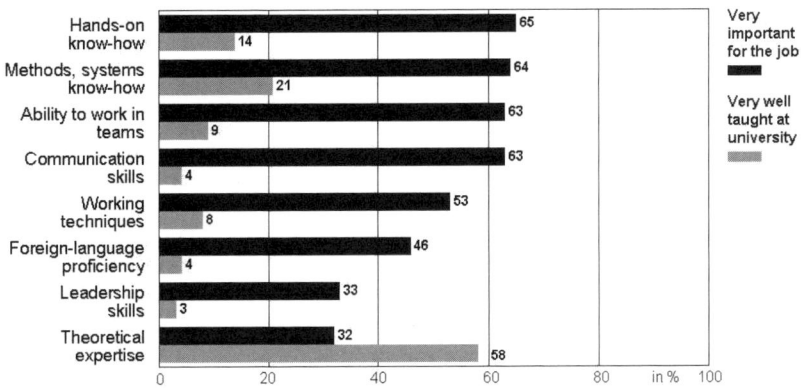

Source: VDE 2009

A similar effect has been observed in other countries. In a much-discussed statement, Lars Pallesen, rector of the highly respected Danish Technical University in Copenhagen, called talks of an engineer shortage *"just talk"* (klynk) on the grounds that otherwise the salaries of Danish engineers would be higher (Pallesen 2007). In Switzerland, Elisabeth Baume-Schneider, Secretary of State for Education, Culture and Sport, stated in 2007: *"Fears about a shortage of engineers have been expressed for many years already but, as far as I know, engineers' wages haven't been increasing at a rate that might confirm that alleged scarcity ..."* (Goossens 2007). In the U.S., a study of the period from 1970 to 2000 produced the sobering result that *"altogether the data ... do not portray the kind of vigorous employment and earnings prospects that would be expected to draw increasing numbers of bright and informed young people into science and engineering fields"* (Butz 2004). Therefore, the European Union warned that *"scientists – especially young ones – need better salaries. ...New human resources for S&T will not be attracted at the required level unless governments translate their political goals urgently into new research jobs and better career perspectives"* (Gago 2004).

Figure 7: Development of entry-level salaries for different groups of engineers in Germany. No "shortage bonus" can be detected.

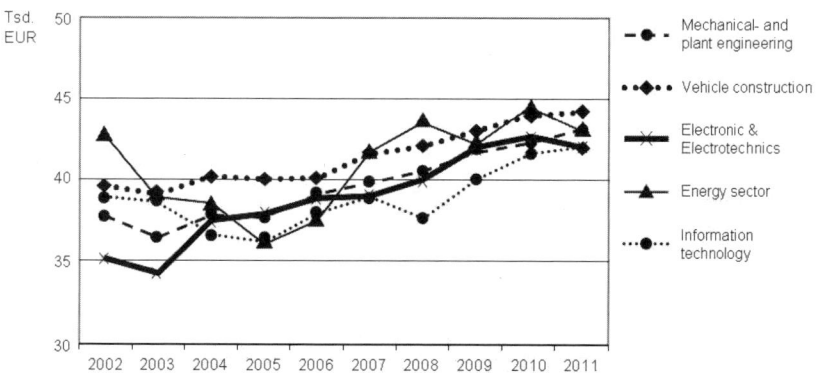

Source: VDI 2011

Of course, even in a market economy, there is no strict correlation between a shortage of applicants and salary increases (otherwise the exorbitant rise in managing board pay in the past would point to a serious lack of capable candidates …). Here, factors like mindset, training and negotiating power play an important role. But such power depends to a large extent on the background of the top decision-makers. And here too, it doesn't look good for engineers, for whom seniority is seldom a bonus and who are not optimally prepared for a management career by their educational background.

The management skills mismatch shown in Figure 8 has to be taken seriously, since PhDs are an absolutely elite group, which even in a company like Siemens (Germany) account for <3 percent of the total workforce. If not offset, these deficiencies can turn into serious career obstacles. Comparable findings have been reported in other countries (e.g. Nair *et al.* 2009; Becker 2011a) and point to a general problem in engineering education. Studies indicate that engineers are willing to become managers (Schmauder 2004; Universum 2009). They are, however, more motivated by their distrust of the career opportunities available to them as technical specialists (Schmauder 2004) than by a drive to leave their world of solid, calculable facts and to move into an area more determined by human interaction, economic necessities and personal vanity (Reisach 2010), for which their educational background did not prepare them well.

Figure 8: Comparison of skills expected by employers in the machine-tool industry and their evaluation of the actual performance of PhD graduates

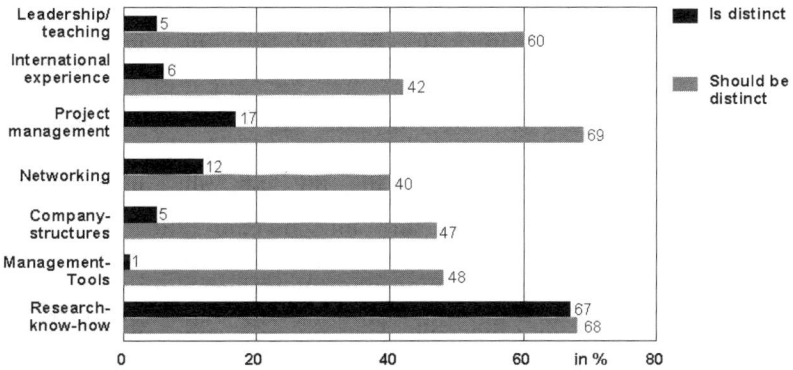

Source: VDMA 2007

Figure 9: Decline in the percentage of managers with a technical background vs. rank at Siemens Germany, 8/2008

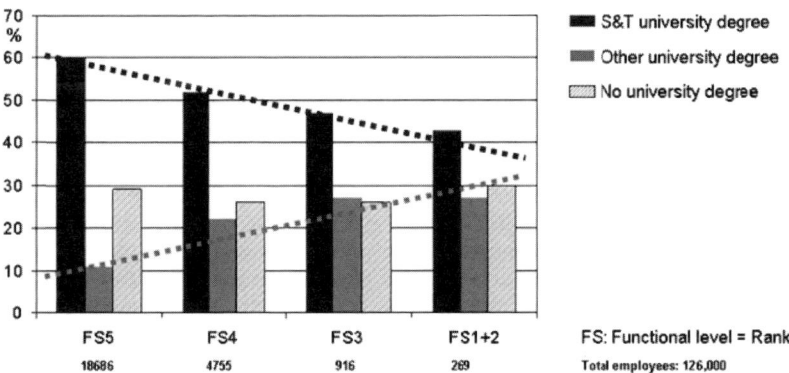

Source: Becker 2008

Consequently, even in a technologically oriented company like Siemens, in which the environment is surely favorable, training as an engineer or scientist is far from being the best career track to a top position. This is clearly indicated by the relationship – shown in Figure 9 – between the educational background of Siemens managers and their relative position in the company hierarchy.

This trend, which increases as one moves upwards, has become stronger over time. In 2011, only 40 percent of Siemens' Managing Board members were scientists and engineers – in 2001, this figure was still 64 percent.

In small and medium-sized enterprises, the situation is different, but the big companies are the showcase and offer the highest salaries (VDI 2011).

The "best and brightest" young minds – who are always and everywhere in demand – are well able to calculate and judge which type of education will lead them to the top positions in companies – and in society.

6 Image and status in society

The changing attitude of society toward technology mentioned in section 2 has had a negative impact on the image of engineers. In a survey conducted in 1971 by the German sociologist Eugen Kogon, 72 percent of some 25,000 responding engineers agreed grumblingly to the statement that "*engineers are the camels on which businessmen and politicians ride*" (Kogon 1971). This situation has not significantly improved – in spite of the fact that technology is now omnipresent. Young users are fascinated by it, but this does not translate into a high attractiveness towards the profession that creates all these devices (NaBaTech 2009). As Figure 10 shows, there is a large gap between the positive self-perception of engineers and the external view – except with respect to good career prospects (VDE 2007). Values in brackets give the results of the 2003 survey.

In general – not only for young women – society fails to provide sufficient visible "role models" of people who have succeeded as engineers rather than by switching from engineering to another profession. As a matter of fact, we live in a media-dominated "jackpot society" in which the apparent success of a few – albeit highly unlikely – "winners" motivates considerably more young people to invest in such "lottery ticket careers" than in less thrilling, but more secure alternatives. It is a safe bet that the normal engineer earns more than the average pop musician or actor, but there are actors or musicians who earn tens of millions of dollars a year! Even if money is not everything, high salaries and the fame and media attention that comes with them lend glamour to an entire profession. Engineers, on the other hand, do not figure in the mass media and are portrayed at best as harmless eccentrics like the comic figure Gyro Gearloose. Also, the often remote location of typical engineering jobs is considered unattractive (Alpay 2008).

Finally, society also applies different criteria in the area of "prominence" and intellectual property rights – criteria that clearly discriminate against creativity in technological fields. Table 1 lists some of the factors that, curiously enough, are frequently overlooked in the public debate.

Figure 10: Discrepancy between the positive perception of young engineers and the more negative view of the outside world.

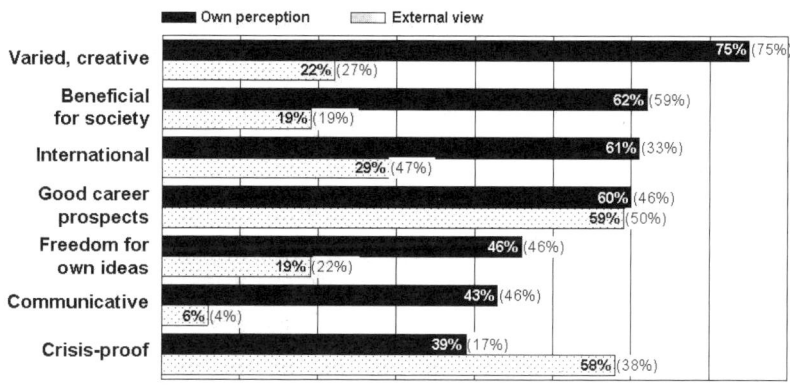

Source: VDE 2007

Table 1: Different yardsticks in society for rewarding intellectual creativity.

Texts and music	**Technical inventions**
To register: easy, cost-free (in Germany VG Wort, GEMA)	To get a patent: expensive, difficult (Patent attorney, patent office)
To maintain rights: free of charge	To maintain rights: rising costs!
Fees collected up to 70 years after death of author.	Maximum duration: ca. 20 years after patent granted, royalties only if used
GEMA: actively protecting authors rights and collecting royalties	Patent Office: No action taken to support inventors

Facts to think about ...
- *Who got famous and made the millions – the inventor of the CD or pop stars like The Rolling Stones, Madonna, Lady Gaga?*
- *Who could veto the modification of the Munich Olympic stadium 30 years after the games – the civil engineer who constructed it or the architect?*

Source: Own table

This disparity between the proclaimed goals and the reality of the social environment sends signals that young people understand perfectly well. Seen in this light, their reaction is a sad, but logical consequence, not an erroneous decision based on a lack of "good advice."

7   Women – aliens in engineering

Engineering fields are traditionally dominated by men – but is this a law of nature? Women in Germany were first admitted to universities only about a century ago, but are now very well represented, accounting for more than 50 percent of total enrolment. They have gained strong positions even in a number of scientific subjects, as shown in Figure 11, based on the 2010 data.

Figure 11: Percentage of women in different scientific majors at German universities.

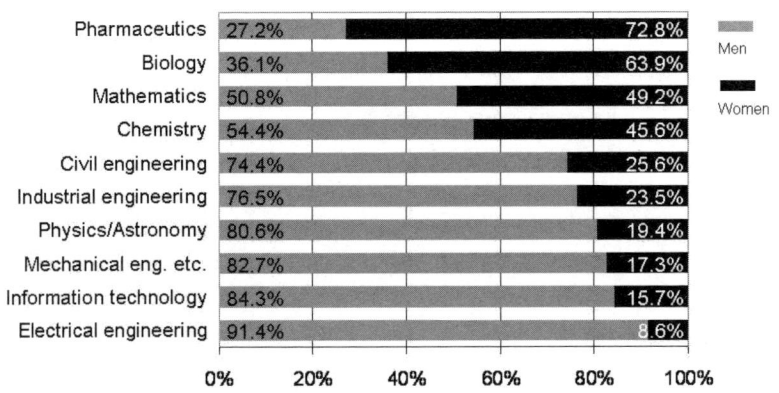

Source: VDI

However, as soon as the word "engineering" appears their share drops dramatically, an effect known from other countries as well (e.g. Eisen 2009). But all attempts to explain by "(VDI)facts" why young women continue to avoid electrical engineering, for example, (in spite of all initiatives to attract them) have been quite unsuccessful.

The only conclusive observation is that women go where they feel at home (Gräßle 2009; Eisen 2009) – that is, where they have overcome a critical threshold, which is probably about 30 percent. After all, the often-cited incompatibility of work and family applies just as much to careers in chemistry (not to speak of law, advertising or medicine, where irregular hours are the norm). And in view of the fact that women now account for half of all students of mathematics (VDI), it is not the rigors of this discipline that are the problem either (Gräßle 2009).

As Figure 2 shows, the unwillingness of young women to become engineers is even more pronounced in highly developed countries (Sjøberg 2010). The more "emancipated" a society is and the greater the range of alternatives that a highly differentiated labor market offers young women, the less they will be inclined to opt for professions they don't wish to identify with. Since this is far too complex a matter to deal with within the limited space of this article, it merely remains to be said that particular sensitivity and support is necessary here – from a psychological rather than an expertise point of view! (For an excellent review, see Hill 2010). The necessary change of mindset must be initiated at school, but must be effected above all in university engineering departments.

Figure 12: First-year students enrolled in electrical engineering in Germany: Progress of women is incremental

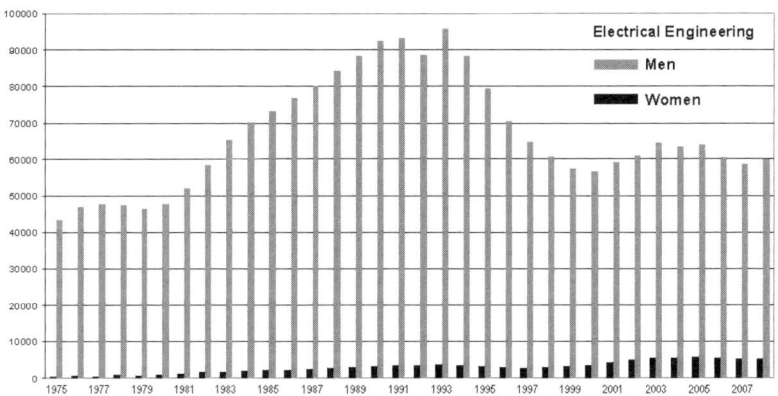

Source: VDI

In addition, the power of perception among young people, which translates into recognition and status in their same-age peer group, cannot be overestimated. Especially during adolescence, self-to-prototype matching is of crucial importance (e.g. Kessels 2005; Taconis 2008; Gräßle 2009; Sjøberg 2010). When defining their gender identity, girls shy away from "male" jobs and vice versa. This classification is strongly determined by traditions in society, prejudices of parents and teachers (Gras-Velazquez 2009) and the perceived image of a profession, which is shaped largely by often gender-stereotypical youth magazines or soap operas (Dahmen 2009). In TV series or journals preferred by young women, prominent figures are lawyers, fashion designers or doctors, whereas engineers simply do not exist. Thus, young women choose alternative careers (Brinck 2008; Hofmeister 2009; Schwarze 2007) – often based on too rosy expectations.

In light of the fact that women are the largest untapped potential for S&T (cf Fig. 12), "changing this in the short term would require nothing less than a cultural revolution. Not changing it as soon as it is possible would be a catastrophic loss for humanity" (Tengelin 2009).

8 What do young people really want?

In this final section, I want to complement my analysis with the results of a recently published large-scale survey of 13,000 young people for which I served on the advisory council (NaBaTech 2009). The scope of the study was to gain a more solid database about the attitudes of schoolchildren (3,006 respondents), university students (6,273) and engineers/scientists (3,470) regarding science and technology. As a general rule, it can be said that their attitude is positive, even if no specific technology is preferred – a finding in line with other studies (Sjøberg 2010). But *"there is a sharp difference between the positive opinion of young people towards science and technology and their actual wish to pursue careers in science and technology"* (OECD 2008). The younger generation's positive attitude doesn't translate into a willingness to make technology their calling. Only 11 percent of the students responding wanted to become engineers, another 8 percent opted for the natural sciences – for the simple reason that the qualities attributed to these professions do no match their wishes, as shown in Figures 13 and 14, which are based on about 2,500 answers.

The mismatch is especially pronounced (and disturbing) as regards the question of a "safe job" – in spite of the fact that students had been well aware of the positive trend on the labor market during the years 2006-2008! But obviously this good news as well as the predictions of future shortages of engineers was unable to compensate for the bad news of the past. Another reason might be the debate about offshoring lower level (= possible entry-level) technology jobs to countries like China and India where cheaper engineers abound. In addition to this distrust, salaries and career prospects were considered critical as well – a finding in line with the results discussed in section 5.

But even important "soft factors" desired in an ideal job shown in Figure 14 differ from those ascribed to technical jobs. As these doubts are obviously to a large extent unfounded (and other professions probably seen in too positive a light), this calls for targeted information campaigns – especially ones presenting role models – to improve the situation.

Figure 13: Profile of an "ideal job" versus a "technical job"

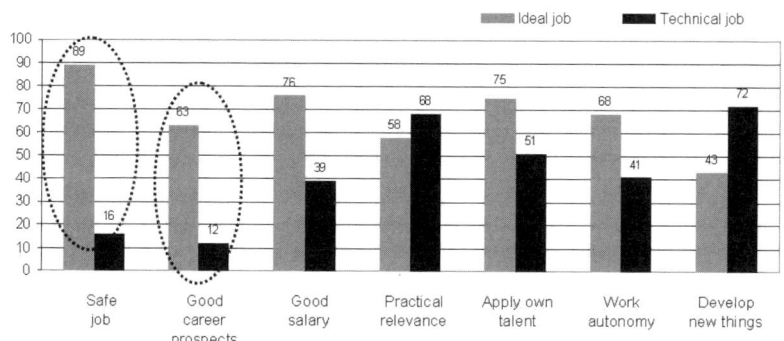

Source: NaBaTech 2009

Figure 14: Comparison of qualities desired in an ideal job compared to a technical job

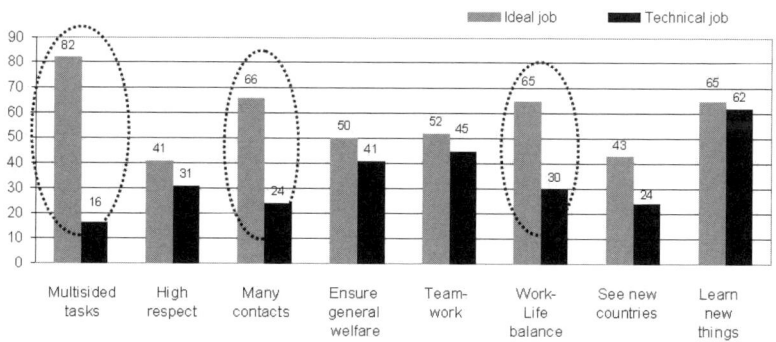

Source: NaBaTech 2009

In this context, some other findings of the NaBaTech-survey may also be mentioned:

- There has been a strong shift in the way children experience technology for the first time. Toys like trains or construction kits (Märklin, Lego, Fischer-Technik) have been replaced by "virtual reality" gadgets like Playstations, the Wii or computer games. This fosters a "just use-it attitude" and increases the importance of technology education at schools.
- Even young women with good grades in math or science are less convinced that they can succeed in engineering – self-doubts that are rein-

forced by the often arrogant attitude of young men choosing technical subjects for the lack of other options.
- Young women starting engineering are not afraid of being a minority, but two-thirds report discrimination afterwards.
- Young men choosing engineering are fascinated by hands-on technology, but are frustrated by theory-laden and often highly abstract initial study phases.

Finally, these findings are supplemented by the results of a recent survey among about 4.200 engineering students. It can be seen that for both sexes a good work-life balance and a safe job are high priorities.

Figure 15: Priorities of German engineering students (3213 male, 1035 female)

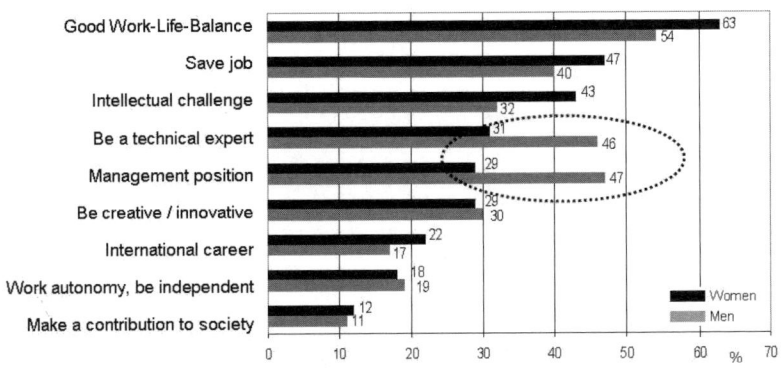

Source: Universum 2009

The biggest gender difference can be found in career goals, as young men display a distinctly higher eagerness either to become technical experts or to pursue management careers. Interesting and important for corporate recruiters may be the fact that an "international career" has a low priority – an attitude corroborated by other studies (e.g. VDE 2009) where 47 percent refused categorically to work for a time in, for example, Asia. Noteworthy as well is the low motivation even of young women to contribute to society, which seems to confirm the "technology hermit" stereotype.

9   Conclusion: What can (and should) be done?

The previous analysis supports the conclusion that young people like technology and do not shun engineering careers just because of laziness or ignorance – they simply don't see it as attractive enough compared to other options. Unfortunately, society and the business world send a host of psychological and financial signals that contradict their claims to foster S&T.

Other trends are contributing to a reduction in the potential supply (OECD 2008): The number of students from blue-collar families, a traditional source of upwardly mobile engineering students, is declining (Hartmann 2009). In addition, the high divorce rates result in more and more children being brought up primarily by their mothers. Since these usually have noticeably less affinity for technology, also through their own upbringing, the children are less motivated to take an interest in technology.

The problem is being further exacerbated by changes in attitude within certain segments of society. Although many of the competition-oriented young people have good grades in science, they prefer to embark on careers in insurance or consulting, where the potential rewards are much more tempting. Young people with a post-materialistic attitude, on the other hand, who value the conservation of nature, are drawn more toward biology or the social sciences. This group considers "machine technology" not as a solution, but as the cause of our problems.

Finally, relatively unattractive teaching methods and a university curriculum that requires students "*with the persistence of Sisyphus and the patience of Job*," as Shirley M. Tilghman, President of Princeton University, characterized it (Tilghman 2010), frustrates multifaceted students, especially young women. In this environment, the younger generation tries to make the choice which best matches their self-perception – and decides against S&T.

To change this, the deterrents analyzed above could be categorized as follows:

Technology – can a commodity become a challenge again?

Technological performance today is determined by "invisible" components like microelectronics and software, has invaded nearly every aspect of our life and has become easy to use. Thus, it is seen as a commodity instead of a challenge – a trend nearly impossible to reverse. As most young people lack the "tinkering experience" at home, schools have a high responsibility as the first (and often only) forum in which hands-on experience with technology is possible. Recommendations:

- Technology education should be a part of every school curriculum. This requires well-prepared (and paid!) technology teachers, interesting and topical material and continuity – starting in kindergarten and not only after adolescence.
- Guest lectures by scientists and industry researchers could bridge the gap between the school curriculum and the fascinating advances of S&T dis-

cussed in the media or seen as products (cars, mobile phones, computers).

S&T and society: Basic knowhow for informed citizens and a part of every one's education

As our life is determined by technology, science education should not just aim at producing a few highly qualified experts, but

"is also about instilling a comprehension of the scientific methods in those who will never oversee a laboratory and giving them a full appreciation of the transformative role of science and technology in daily life. Without well-informed policymakers and a discriminating public, scientific progress will be slowed or misdirected, to everyone's detriment" (Tilghman 2010).

S&T has to become again an integral part of the knowledge of every educated person. Recommendations:

- Everyone teaching S&T – from teachers to scientists – should aspire to make these subjects comprehensible, present solutions in a societal context and be willing to engage in discussion.
- Media – from school textbooks to TV – should not present S&T as some remote and male-dominated topic for specialists, but as the foundation of our well-being, ignorance of which is a severe deficit, not just a pardonable weakness. One way forward would be to create, for example, a "Jules Verne Award" for the best novel with a technology background.
- Intellectual property rights for technological innovations should no longer be discriminated against compared to those for music and texts (for example, through tax concessions).

University education: From stifling to stimulating interest!

For at least four decades (as the author can testify), S&T university professors, in particular, have lamented the inadequate preparation of freshman students and likened their own job to that of erecting pile dwellings in a swamp of ignorance. As a consequence, especially the first study phase has been devoted to laying the "theoretical foundations" and to "weeding out the weak" (Becker 2011b). This approach is an effective selection mechanism, but surely not a means to stimulate interest in the subject and prevent those students who doubt their ability to master engineering from dropping out. Extensive studies have shown that a more hands-on approach can yield much better results (Hake 1998), but such findings have remained without impact on most curricula. Recommendations:

- Shift the focus from even more fundamental research in engineering education to implementation strategies. The top priorities should be best-practice sharing and a subsequent emulation of successful initiatives like the learning factory (Morell 2007) and project-based learning of the type implemented by Professor Hampe at Darmstadt Technical University (Wolf 2006), which involves freshmen in practical design work.
- Stop emphasizing the extremely challenging nature of S&T and the high threshold that must be crossed to achieve an accepted degree. Shift to encouraging and supporting (for example, by introductory courses) the less gifted and less well-prepared students and train them for the many career possibilities available in addition to technical specialization.

Industry: Enable contacts, highlight the opportunities and improve the perspectives

Companies are places where technology is developed and implemented as well as future employers. They play a crucial role in inspiring the interest of the younger generation and shaping the image of the engineering profession. Recommendations:

- Dedicate resources to make it possible for students to come in contact with practical work (factory visits, internships, student programs, presentations by trained experts at schools and universities)
- Highlight the cooperative, problem-solving attitude in engineering, which is a good starting point for graduates to develop their potential. Stop emphasizing that only the best are welcome, as most jobs require "normal good" people. Above all, refrain from pointing out that hundreds of thousands of Asian engineering graduates are eager to work for a fraction of a European salary since the logical conclusion of any intelligent youngster can only be to avoid such an endangered profession!
- Present role models to exemplify the many positions open to technical graduates and the attractiveness of the work, improve the career options for engineers beyond the age of 50 by offering continuous education and challenging projects instead of early retirement.

There is no guarantee that these measures, if applied, will solve the problem, but we can be pretty sure that without them the trend away from S&T will continue – and that as a consequence our developed societies will be unable to master the challenges of the future.

This article is an extended version of the invited talk "**Are young people lazy, blind or misguided? A fresh look at possible reasons for not choosing a career in science and technology**" presented at the Motivation Conference, Dec. 2009, Wuppertal. It is an updated version of an article published in the *European Journal Engineering Education* (2010), Volume 35 Issue 4: 349. The *European Journal Engineering Education* is available online at: http://www.tandfonline.com/ceee.

References (most internet sites were accessed in January 2010)

Abbot, A. 2003. The Zen of Education, http://magazine.uchicago.edu0310/features/zen.shtm

Alpay et al. 2008. Student enthusiasm for engineering: charting changes in student aspirations and motivation; European Journal of Engineering Education 33 (5-6): 573-585

Becker, F. S. 2006. Globalization, curricula reform and the consequences for engineers working in an international company, European Journal of Engineering Education 31 (3): 261-272

Becker, F.S. 2008: Scientists in demand. What is keeping students from pursuing careers in science? Presentation at the Lisbon Council 2008 Skills and Human Capital Summit; Brussels September 16, 2008; http://www.lisboncouncil.net/initiatives/human-capital.html

Becker, F.S. 2009. Why not opt for a career in Science and Technology? An analysis of potentially valid reasons; paper 131, Proceedings of the 37h SEFI Annual Conference, Rotterdam 2009; ISBN 978-2-87352-001-4

Becker, F.S. 2010. Why don't young people want to become engineers? Rational reasons for disappointing decisions; Europ. Journal of Eng. Education Vol. 35, No. 4, August 2010: 349-366; http://www.stepstwo.ua.ac.be/~stepstwo/Becker_EJEE_Why_dont_young_people_August-2010.pdf

Becker, F.S. 2011a. Graduate employability – what is needed and how employers and universities can cooperate to achieve it; presentation at the Scottish Higher Education Employability Conference; Edinburgh, June 1st, 2011; http://www.heacademy.ac.uk/events/detail/2011/jointevents/01-02_June_2011_Scottish_ Employability_conference

Becker, F.S. 2011b. Quality in Engineering Education – an Industry View; SEFI Annual Conference/1st World Engineering Education Flash Week; Lisbon, Sept., 28-30, 2011, paper No. 10

Becker, F.S. 2011c. Berufsfähigkeit als Herausforderung für die Hochschulbildung. Anforderungen an Berufseinsteiger und Karrierebewusste aus Sicht der Industrie. Vortrag an der HTW Dresden, Oct. 25th, 2011; http://www.htw-dresden.de/fileadmin/userfiles/htw/docs/Studium/251011_SI_Becker_HTW.pdf

Becker, F.S. 2012a. Qualität der Ingenieurausbildung – Betrachtungen aus Industriesicht; http://www.zvei.org/Verband/Publikationen/Seiten/Qualitaet-der-Ingenieurausbildung---Betrachtungen-aus-Industriesicht.aspx

Becker, F.S. 2012b. „Herausforderungen für Elektroingenieure/innen"; http://www.zvei.org/Verband/Publikationen/Seiten/Herausforderungen-fuer-Elektroingenieure-innen.aspx

Brinck, C. 2008. Die Freiheit, sich gegen den Ruhm zu entscheiden; in: Frankfurter Allgemeine Sonntagszeitung, 08.06.2008

Butz, W. P. et al. 2004. Is there a Shortage of Scientists and Engineers? How would we know? http://www.rand.org/pubs/issue_papers/IP241/index.html
Dahmen J.; Thaler A. 2009. Image is everything – is image everything? Proceedings of the 37h SEFI Annual Conference, Rotterdam; ISBN 978-2-87352-001-4
Eisen, B. 2009. http://www.insidehighered.com/news/2009/07/22/stem
EU 2006. European Commission: Women in Science and Technology – the Business Perspective; Report EUR 22065 EN; http://www.businessupdated.com/show news.asp?news_id=1354&cat=Women+in+science+and+technology:+the+busine ss+perspective
EU 2008. Flash Eurobarometer Series 239, Young People and Science, European Commission
Fabian, G.; Briedis, K. 2009. Aufgestiegen und erfolgreich; Ergebnisse der dritten HIS-Absolventenbefragung; HIS-Forum Hochschule, 2/2009: 65
4Ing 2009. H. Viele soziale Aufsteiger unter den Professoren der Ingenieurwissenschaften und der Informatik; http://www.4ing-online.de/fileadmin/uploads/ presse/20100104.pdf
Gago, J.M. 2004. Europe needs more scientists; report by the High Level Group on Increasing Human Resources for Science and Technology in Europe; ISBN 92-894-8458-6
Gerlach, G. 2007. Die Situation ist in der Tat bizarr, Interview in „Markt und Technik", 30.03.2007: 52
Goossens, M. 2007. Where is the shortage of Engineers in Europe coming from? A historical perspective, SEII Round Table Conference Nov. 9/10, 2007, Official report: 23-31; http://www.nuttin.info/SEII/doc/Report-TRI.pdf
Gräßle, K. 2009. Frau Dr. Ing. – Wege ebnen für Frauen in technische Studiengänge, Leverkusen, ISBN 386649243X
Gras-Velazquez, A. et al. 2009. White paper women and ICT, European Schoolnet; http://seskills.eun.org
Hake, R. 1998. Interactive-engagement vs. traditional methods: A 6000 student survey of mechanics test data for introductory physics courses. American Journal of Physics 66(1): 64-74
Hartmann, M. 2009. Stellen die Ingenieurwissenschaften noch den Karriereweg für soziale Aufsteiger dar? In: Nagl, M. et al. (Hrsg.) Zukunft Ingenieurwissenschaften – Zukunft Deutschland; Springer Verlag: Berlin: 191-199
Henning, K. 2007. Bachelor nach rheinischer Art, Die Zeit, 25.10.2007: 82
Herrmann. 2010. Kampf um eine Marke, Süddeutsche Zeitung, July 5, 2010: 52
Herrmann. 2011. Oral statement made at the "VII. Symposium Hochschulreform", Munich, April 1
Hill, C.; Corbett, C.; St. Rose, A. 2010: Why so few? Women in Science, technology, Engineering and Mathematics; AAUW Research Report; http://www. aauw.org/learn/research/whysofew.cfm
Hofmeister, H. 2009. Warum verzichten wir auf 40% unserer Kreativen? In: Nagl, M. et al. (Hrsg.) Zukunft Ingenieurwissenschaften – Zukunft Deutschland; Springer Verlag: Berlin: 177-189
IW. 2008: Ingenieurmangel: Übel an der Wurzel packen; iwd Nr. 9, 2008 http://www.iwkoeln.de/tabID/2204/ItemID/21933/language/de-DE/default.aspx
Johnson, W. C. & Jones, R. C. 2006. Declining Interest in Engineering Studies at Time of Increased Business Need, in: Weber, L.E & Duderstadt, J. J., Universi-

ties and Business: Partnering for the Knowledge Society, Economica, London: 243-252
Kessels, U. 2005. Fitting into the stereotype: How gender-stereotyped perceptions of prototypic peers relate to liking for school subjects; European Journal of Psychology of Education XX (3): 309-323
Kogon, E. 1976. Die Stunde der Ingenieure, VDI-Verlag: Düsseldorf
Meyer, T. 2008. Deutsche Bank Research: MINT-Fachkräfte – Zwischen zyklischem Engpass und Strukturwandel; Report, 16.07.2008
Minks, K.-H. 2005. Kompetenzen für den globalen Arbeitsmarkt: Was wird vermittelt? Was wird vermisst?, in: Grünberg/Wenke (Hrsg.), Arbeitsmarkt Elektrotechnik Informationstechnik 2005, ISBN-13: 978-3800728961: 29-48,
Minks, K.-H. & Fischer, L. 2007. Acht Jahre nach Bologna- Professoren ziehen Bilanz, HIS Projektbericht, 10/2007
MoMoTech. 2009. http://www.dialogik-expert.de/en/forschung/projektverbund_zukunft.htm
Morell, L. 2007. Engineering Education, Globalization and Economic Development: Capacity Building for Global Prosperity; World Innovations in Engineering Education and Research; W. Aung et al. (Ed.), iNEER, Begell House Publishing, New York: 25-43
NaBaTech. 2009. http://www.acatech.de/de/projekte/abgeschlossene-projekte/nach wuchsbarometer-technikwissenschaften-nabatech.html
Nair, C. S. et al. 2009: Re-engineering graduate skills – a case study, European Journal of Engineering Education 34 (2): 131-139
OECD. 2007. Report Education at a Glance, ISBN 978-9264032873
OECD. 2008. Encouraging student interest in science and technology studies; ISBN 978-9264040694
Pallesen. 2007: http://ing.dk/artikel/77777-dtu-rektor-ingenioermangel-er-klynk
Prieto, E. et al. 2009. Influences on Engineering enrolments. A synthesis of the findings of recent reports, European Journal of Engineering Education 34 (2): 183-203
Pritschow, G. 2004. Herausforderung als Chance; Tagungsband zum acatech-Symposium "Innovationsfähigkeit", Berlin: 11. Mai 2004: 50
RAND. 2004. The US Scientific and technical Workforce; http://www.rand.org/pubs/conf_proceedings/2005/CF194.pdf,
Reisach, U. 2010. Macht und Machtmissbrauch; Personalwirtschaft 2/2010: 22-24
Schmauder, S. 2004. Fachwissen ist und bleibt Basis für die Ingenieurkarriere, Grüneberg/Wenke (Hrsg.), Arbeitsmarkt Elektrotechnik 2004, ISBN-13: 978-3800728992: 69-77
Schwarze, B. 2007. Ingenieurinnen in Studium und Beruf – zwischen Herausforderungen, Stereotypen und Berufsengagement; Das Berufsbild der Ingenieurinnen und Ingenieure im Wandel; VDI-report 37
Sjøberg, S. & Schreiner, C. 2010. The next generation of citizens: attitudes to science among youngsters; in: Bauer, M; Allum, N. &. Shukla R. (Eds), The Culture of Science – How does the Public relate to Science across the Globe? New York, Routledge, in print
Staudt, E. 1998. Strukturwandel und Karriereplanung, Herausforderungen für Ingenieure und Naturwissenschaftler; Springer Verlag (1998): 4

Taconis, R. & Kessels, U. 2008. How Choosing Science depends on Student's Individual Fit to 'Science Culture'; International Journal of Science Education, May 2008: 1115-1132

Tengelin, N. 2009. "Corporate Investment to Increase Interest in Mathematics, Science and Technology; A general analysis for the Volvo-group"; Master Thesis, Chalmers University Sweden, Report No. E 2009: 093

Tilghman, S. M. 2010. The Future of Science Education in the Liberal Arts College; speech given as president of Princeton University at the Presidents' Institute annual meeting in Florida, Jan 5

TU9. 2009. Press release, Dec. 8, http://www.tu9.de/presse/3310.php

Universum. 2009. Universum Student survey

VDE. 2000. Studie Ingenieure der Elektro- und Informationstechnik;

VDE. 2007. Studie "Young Professionals 2007" (in German)

VDE. 2009. Studie „Young Professionals 2009" (in German)

VDI. 2010. Ingenieure haben weniger Kaufkraft als vor 10 Jahren; VDI nachrichten, 15.01.2010

VDI. 2011. VDI nachrichten Studie Ingenieureinkommen 2011. In: *Germany, the salaries of standard wage earners working in the thriving Energy sector, if adjusted for inflation, showed a slight decrease between 1997 and 2007!*

VDI. Extensive data to be found at http://www.vdi.de/wirtschaft-politik/arbeitsmarkt/monitoring-datenbank/#hochschule (in German)

VDMA. 2007. Impuls survey „Anforderungen an die Promotion im MB und der Verfahrenstechnik", Frankfurt, www.impuls-stiftung.de

VDMA. 2009. Impuls survey, Zwischen Studienerwartungen und Studienwirklichkeit – Gründe für den Studienabbruch. http://www.his.de/pdf/21/studienabbruch_ursachen.pdf

Winckler, G. & Fieder, M. 2006. Declining Demand among Students for Science and Engineering?, in: Weber, L. E.; Duderstadt, J. J., Universities and Business: Partnering for the Knowledge Society, Economica: London: 233-241

Winkelman, P. 2009. Perceptions of mathematics in engineering; Europ. Journal of Eng. Education Vol. 34, No. 4, August: 305-316

Winkler, H. et al. 1996: VDI-Studie Ingenieurbedarf, Universität Gesamthochschule Kassel, 30.06.1996

Wolf, S. & Hampe, J.M. 2006. How to provide first-year students with a really good start to their study program; Proceedings of the ASEE Annual Conference, Chicago, June

ZVEI. 1987. Erfahrungsbericht des ZVEI auf der CeBit-Messe, Protokoll vom 12.10.1987

ZVEI. https://www.zvei.org/de/forschung_bildung/ingenieurausbildung/statistik/

Mind the gap: Science and young people

Federica Manzoli, Flora Di Martino, Daniele Gouthier, Donato Ramani

Abstract

It is well known that in Europe young people are losing contact with science and the number of science students in universities has decreased. At the same time, in this context a gender difference exists: Girls appear less interested in science, engineering and technology (SET) than boys.

In order to understand what girls and boys think about the SET professions and what is the present situation in terms of gender difference, a research-action experience was developed, as part of the European project GAPP (Gender Awareness Participation Process). A sample of high school students and professionals engaged in academic or other science-related careers were involved and interviewed as the first step of the *process*. Following a bottom-up approach, on the basis of these first results and of six Open Space Technology activities held in the partner countries, a set of successful pilot activities were realized. Aim: Bringing young people, and especially girls, closer to SET professions.

1 Bringing young people closer to science and technology professions

In recent years, many studies in the field of education and sociology affirmed the importance of monitoring the relationship between young people and science in order to bring them closer to professions related to science, technology and engineering. The most recent surveys (European Commission 2008) on this topic show that while youngsters declare a high level of interest towards issues regarding science and technology, they are scarcely willing to enterprise a scientific career. These data reveal the existence of a deep gap between SET perceived through the media, the school and the civil society as a whole and poses profound questions to the school systems of many European countries (Gouthier and Manzoli 2008; Schreiner & Sjøberg 2006).

From this question started the European project GAPP – Gender Awareness Participation Process, funded under the 6th Framework Programme and carried out in the years 2007-2008. The results of its different phases are discussed in this[37]. The premise of the project was that, in order to develop an effective information and education action, it is needed to apply a *bottom-up approach*. To start from the ideas, opinions and attitudes expressed from the targeted publics, the communication activities can be planned and implemented in a more effective way.

The project consisted of two main parts: The first dedicated to carrying out a qualitative research on different publics: Young people aged 14-18, their parents, teachers of high schools and professionals working in the field of science and technology. The second part aimed on developing concrete activities for bringing young people closer to SET careers.

Subsequent to the first phase of the project, the operators involved realised that the initial goal of favouring the choice of a scientific career among young people had to be replaced by the goal of making them aware of who the people working in SET are and what they do, the first step needed to spur them to embark on this type of career.

Then, awareness-raising became the central goal, to be achieved through concrete practices, to be participatory and appealing to young people.

Based on a "philosophy of practice", the six GAPP-involved countries developed a series of Pilot Activities (PA). Each country devised and implemented different participatory projects, very various and rich in ideas on their future implementation. Thus, the result was a project with a real European scope, given the common goals and the range of propositions.

The work was carried out in the framework of the science centres, which contacted and collaborated with local schools; the result was the identification of new and viable methods to make science centres and schools work together, directly involving the world of research.

2 Methodology

The different stages of the project requested different methods, which have been successfully integrated, starting from what the targeted publics think of the scientific careers and arriving to the implementation of the pilot activities.

---

37 The participants of the project were four science centers and three scientific institutions: Fondazione Idis, City of Science, Naples (Italy); Sissa, International School of Advanced Studies, Trieste, (Italy); Youth Research Centre, University of Warsaw (Poland); Nemo Science Center, Amsterdam (The Netherlands); Royal Belgian Institute of Natural Sciences, Brussels (Belgium); Experimentarium, Copenhagen (Denmark); Ciencia Viva Science Centre, Lisbon (Portugal).

2.1 The qualitative research phase

In order to assure uniformity in researching the perception of the SET careers in six different European countries, with six different school systems, particular attention was paid to the design of the study. The shared choice was to involve all the targets of the project: *high school students, teachers, parents*.

The method chosen was that of the focus group, as mean of carrying out a genuine research-action.

In any country the 8 focus groups were targeted as follows: 2 with participants aged 14-16; 2 with participants aged 16-18; 2 with science teachers; and 2 with parents having at least one son or one daughter aged 14 to 18.

After a comparative analysis of the different school systems, the age range of the young people to be involved in the focus groups was between 14 and 18. In order to make sure of the right performance of the targeted groups, without mixing adolescents in a too different stage of life, teenagers in the lower range (more or less 14-16) were separated from those in the higher range (more or less 16-18). Each individual country was free to arrange the ages according to the time when students choose the specialization after the compulsory school, which is very variable from country to country (from 13 to 18).

The science teachers were recruited from different types of schools (technical, professional, gymnasium, etc.), among the core disciplines of the project (mathematics, chemistry, IT, technical subjects, etc.). Either for teachers and parents, the commitment of the recruiters was not to mix up students, parents and teachers related to each other (i.e. not involving students in the same class of the teachers, parents of different students as the participants to the focus groups).

All the targets were selected according to shared screening questionnaires.

Partners were provided with a common guideline, topics of which were the basis for the national reports and allowed the comparison among countries. The focus group discussions were transcribed and the analysis of the transcripts processed through the discourse analysis (Gaskell and Bauer 2000; Maxwell 2005).

The main issues of the guideline were: A map of the terms *science* and *technology*, the perception of the SET professions in their country, the proposal of initiatives in order to bring the young people closer to these professions.

The second part of the research was the interviews with a sample of opinion leaders in the scientific and technological world. 10 in-depth interviews per country were realized. Participants involved were academics in SET fields (math, physics, chemistry, biology, IT, engineering, earth sciences), managers in companies related to these disciplines, directors of science parks and science museums, administrators and politicians in the field of SET, experts of gender issues. Little more than a half of the respondents were women.

As for the focus groups, a common guideline was shared, either for the interview and the national reports.

Under a gender perspective, the main questions posed to interviewees regarded: Their own experience in the scientific and technological world of work, their point of view on the present situation (e.g. the distribution of women and men in the organization chart of their institution or company and its meaning), their proposals for increasing girls' interest in SET careers.

The design of the research was created by the scientific partner of the GAPP project (Sissa, Italy), but authors of the results are the partners in the six countries. Their national reports were elaborated in two full reports: One for the focus groups and one for the in-depth interviews[38].

2.2 The open space technology

Coherently to the qualitative research method, each GAPP partner organized an Open Space Technology event following the previous results of the project. The aim of the OST workshops was to obtain, in a participative way, a list of proposals for defining practical pilot activities actions involving scientists, engineers and professionals with SET backgrounds from the public and private sectors to be tested in high schools with students.

Open Space Technology is a self-organizing practice that enables groups of any size to address complex, important issues and accomplish meaningful work. It releases the inherent creativity and leadership in people by inviting them to take responsibility for what they care about. Open Space establishes a *marketplace of inquiry*, where people offer topics of interest, reflect, learn and work together (Howen 2008).

Targets were the same as the previous qualitative research, the number of which was enlarged in order to properly run the activity.

2.3 The evaluation

The task of the scientific partner was to carry out the evaluation process of the project. In order to test either the proceedings of the GAPP activities and their effectiveness, a mixed approach of quantitative and qualitative methods were set up.

---

38 Reports and documents related to the GAPP project can be requested to: federica.manzoli@gmail.com or dmartino@cittàdellascienza.it.

For the former, a questionnaire was submitted to the partner at the conclusion of the project meetings.

For the latter, a questionnaire comprehending closed and open questions was settled and tested by the students, teachers and parents attending the pilot activities.

Furthermore, participant observation by each partner was realized. This latter was conducted through a common grid settled by the scientific partner.

3 Results

3.1 A bottom-up approach: What young people, parents, teachers think about S&T careers

The most significant division appearing in the students' classifications on science from the focus groups discussions is the one between science itself and science understood as education & learning process experienced at school (Gouhier and Manzoli 2008). The first is evaluated positively and is described in terms of *adventure, challenge, satisfaction, progress*. The second is perceived negatively, in terms such as *boredom, hard* and *unfruitful work, and stress*.

An important result is the perception of the relation between science and technology: They do not have a clear idea of the link between the two. Probably because of their school curricula, they tend to keep *science* and *technology* separated, instead of thinking of their total co-existence in the current scientific enterprises. Thus, youngsters show an *ancient* idea of the division between science and technology – actually it disappeared in the past century (Novotny 2005).

In terms of gender difference and attitudes towards different disciplines, we confirm the results of the current studies on this topic: Girls tend to emphasise the social outcomes of SET concerning health issues a little more, whereas boys are slightly more attentive to the success that some well-known male scientists or technological entrepreneurs have achieved.

Similarly, the representation of the scientists conform to the current stereotype: Scientists are half heroes, half crazy people, completely focused on their scientific issues, travelling a lot (which is one of the most important discriminating factors in the gender issue), working alone (in professions as the mathematician or the IT expert). In the common imagery, the representation of a researcher is very much a male one.

On the other hand, when asked about what a scientist does, the main result is a vague idea of the work of science. For example, in the focus groups with teenagers aged 14 to 18, before asking whether they wanted to become mathematicians, they were asked to describe who she or he is and we found out that young people do not have a clear view of what a mathematician does. In the second part of the FGs, we were asked to select some images taken from generic magazines describing the scientific professions among: Engineering, IT, chemistry, mathematics, physics, and biology. Nobody made an attempt at using images to describe the job of a mathematician. Firstly, because it is materially difficult to find images describing this job, but clearly it is also difficult to mentally picture a professional environment where he or she works.

The main influences in the choice of a career are personal interests, teachers and parents. In general, students think the choice is theirs, parents think teachers have the most important influence on their children and teachers believe it is parents who decide in higher social background and friends elsewhere.

In any case, all the targets believe it is very much important to have direct contacts with the SET world during the school period.

3.2 Talking to the experts: What the opinion leaders think about young people and SET professions

The goals of this part of the research were to learn and to understand the interviewees' personal stories, their scientific, cultural and social background and the difficulties they had to face in their career, and their ideas on the difficulties that young people may encounter when engaging in such a career. As the interviewees were chosen within a circle of privileged witnesses, interviewing was also aimed at learning their vision on the future career opportunities for young people of both genders in their field of research or activity.

In most of the biographies the gender issue does not turn up and is not a matter mentioned in itself. The interviewees perceive this discrepancy as natural, enrooted in cultural and sometimes biological differences. All of the interviewees – men and women alike – highlight some aspects in which their gender has always played an active role for their career. Not in all cases are they negative conclusions. Instead, some advantages provided by gender are frequently mentioned. A clear result is that a researcher may or may not be a good researcher, totally irrespective of their gender.

Prejudices are yet even more evident when responsibilities must be shared and distribution of power is concerned, because historically men are in control. A female researcher is first seen as a woman and then as a researcher. According to the interviewees, given a substantial equality in the possibility of becoming a good researcher, there is a different social pressure on male and female workers, male and female scientists, wives and husbands, owing to historical and cultural reasons. This also means that being a woman can also be a further opportunity because it undoubtedly favours visibility. On the other hand, today women are induced to search for an intellectual compensation.

Almost all of the interviewees stress the importance, for the emergence of a scientific vocation, of early contact with science and with practices that could raise the youngsters' curiosity about the world.

The expectations for the change of the quantity and the role of women are first of all connected with civilization changes. Much hope is connected with the growth of economic conditions, population and more funding for science. As for practical solutions, they deal with the reforms in educational systems, however not in structure but in substance. There was also an agreement on policy-based changes, which should target at an equalisation of the proportion of gender in science. The idea of parity was especially criticised. All the interviews seemed to reveal the consciousness of a cultural change with a dual impact. The positive one is about the increasing equality of women, differentiation and exchange of social roles. The negative one centred on the decrease of interest in difficult sciences by young people, not only girls.

3.3   The pilot activities

In the GAPP project, Pilot Activities are implemented to integrate the learning process at school with practical activities that provide students with an insight about "what is behind a scientific career", in terms of meaning, models and diversity of professions connected to science and technology. Therefore, the involvement of school teachers, students and professionals in SET underlies all the activities.

Pilot activities consist in activities that spur young people towards a scientific interdisciplinary view, more capable to meet the complexity of the contemporary world, explaining how science and technology are connected to their everyday life, and giving them a modern vision of science, on the one hand, while exploring the gender gap in science and technology professions. The result starts from the scientific interests of young people for developing hands-on activities, workshops, visits, laboratory activities in different institutions, in order to create a connection between what they studied and the real world.

Each pilot activity features an active meeting of young people with researchers in the field of SET.

3.3.1   Working in scientific laboratories (City of Science, Italy)

On the basis of the need to make young people aware on what science and technology professions are, Pilot Activities were aimed at:

- Connecting the schools with the world of the research;
- building a direct contact between researchers and students;
- new opportunities for students to understand scientific data processing in the business world.

The core of the activities was a series of visits of the students involved in nine laboratories, where the researchers have welcomed them, explained their activity, their everyday life, along with the understanding of specific scientific disciplines.

The organisation of the activities started with meetings between teachers and City of Science staff, and between the researchers and City of Science staff. The teachers emphasised the didactic value of practical activities in the laboratories on scientific and technological subjects connected to scientific and technological topicality, real life and work. Researchers also agreed on these remarks and they proposed activities on basic physics, nanotechnology, seismic risks, nanoelectronics, and the environment: Quality of air, the composite materials, biochemistry, and volcanic risks, to be held in the university laboratories and in the research centres.

The activity chosen was a practical activity in the research laboratories on scientific and technological subjects connected with scientific and technological topicality, real life and work.

Each activity was divided in two main steps:

- The first one was a practical educational activity in the scientific laboratory of the research Institute involved.
- The second one was a meeting with the scientist to discuss directly with him/her the topic of the research, the new applications in the future and to understand the curricula needed to pursue the same career.

All the activities were terminated at the end of the fifth month with a plenary meeting of all participants (with a smaller number of students), where they discussed the experimentation implemented and the future development of further significant activities to be included in school curricula and in Universities as vocational guidance for students.

3.3.2 The scientist's blog (Ciencia Viva Science Center, Portugal)

This activity promoted communication between students and scientists through a blog hosted by Ciencia Viva. Scientists contributed to this blog in three major areas: Work, career and personal areas. The activity also offered a visit to their institutions and laboratories, providing the students with a direct contact with the work done in the institutions by scientists and other researchers.

The first step was the creation of the blog: In order to find a functional blog editor, as far as administration and usage were concerned, a search on blog editors was performed. After choosing the blog editor, it was necessary to publish it, that is, to choose a server to physically allocate the software and to make the editor itself available. After some technical considerations, it was decided to associate the blog to Ciência Viva domain. The blog started being designed from an attractive point of view.

After everything seemed to be correctly functioning, training started with the people who would directly or indirectly be using the blog. So, a mini-manual was created with the explanation of all basic functions, such as: How to create a new item in the blog, how to introduce links for external sites, among some other useful information on editing the blog.

During the conception of the blog's structure and its contents, it was decided to build a main page with links to the blogs of each institution.

All this period of time included: Selection of the blog editor, installation and evaluation in a test server, in order to exclude editors that were not interesting, selection of the host server, installation and configuration of the blog editor, accessibility performance evaluation, graphic layouts, contact with schools and researchers, users and administrators training, creation of the blog structure, accessibility and permissions.

The activity lasted for 4 weeks and it ended in the 5th month with the visit to the scientists' laboratories. As far as planning is concerned, the role played by the teachers in activating and coordinating the students was fundamental.

3.3.3 Science students as role models, dialogue as a method (Experimentarium Science Center, Denmark)

The aim of this activity was to present students and parents with role models and to motivate and engage them in group discussions of science education and career matters, so as to make them aware of what science and technology professions can imply. The activity envisaged meetings at school featuring students and parents with experts from the science centre Experimentarium. The class visits imply a small group of Experimentarium's young students-teachers (called pilots) visiting the classes.

All of these pilots have been in the midst of studying various 5-year scientific curricula. Therefore, the visits gave pupils and parents a chance to be face to face with someone who was studying and had chosen science as a career and entered dialogue together. The visits started with Experimentarium's pilots introducing themselves and their studies. Following this, pupils and parents worked together in three workshops:

Picture lottery

In this workshop, the groups had to match game pieces with people possessing various types of education. Among the game pieces there were young and old, famous as well as unknown people. The message was that scientific studies can open doors to a wide variety of work, which corresponds to the varying types of people, enrolling in the different scientific educational institutes.

Free fall

In this workshop, using very few means, the groups had to construct a device, which could protect an egg, so that it could drop for about a meter without breaking. The message here was that it takes a combination of several different skills to implement such a task, and that some of these skills might not normally be associated with science. This corresponds very well to how many scientific work processes are carried out.

Everything in its own place

In this workshop the group had to divide a selection of everyday products into groups that corresponded to the particular types of education lying 'behind' the products. The message was that if you want to make a difference in the world and develop things that everyone can appreciate, a scientific education might just be the way to do it and that there is a need for varied and different science skills for many products.

3.3.4 Meeting the (female) scientist (Royal Belgian Institute of Natural Sciences, Belgium)

This Pilot activity envisaged the organisation of four meetings and discussions with female scientists and secondary school students. Meeting scientists who are women and sometimes mothers and/or wives could have an impact on girls who otherwise would not have chosen a career in SET, thinking that it would not allow them to lead a career and a family/social life at the same time.

Four pilot-activity meetings between scientists and students were organised: Three of these meetings took place at the Museum of the RBINS, while the fourth one took place in a school classroom. Differently from the first one, the latter saw the presence of three scientists at the same time. Moreover, it was less focused on the scientists' subject of study than the other meetings.

The first step was the recruitment of schools, through the contacts existing between RBINS and local schools.

At the Museum

On the day of the meeting, students and their teacher were received by our GAPP reporter at the entrance of the Museum. The scientist joined the group and guided the participants into the Museum where she practices her discipline. There, she talked for approximately one hour about her work. This presentation was followed by about one-hour explanation of the scientist's job in her office, demonstrating the functioning of the instruments and materials (bones, skulls, computers). By the end of each meeting the person in charge of the activity allowed some a question-answer session between the pupils and the scientist about her career and training.

In the classroom

The GAPP reporters gathered the students in a classroom. They were told they were going to meet three scientists available to talk to them about their jobs and careers and to answer their questions. They did not know the scientific fields that were to be represented and did not know that only women scientists were to participate. Each scientist had taken with her some scientific material (objects, maps) that could represent her discipline. One after another, they showed these objects to students who had to guess what their work was. Once the job was made clear, the students and the scientist launched a discussion about the reasons behind this career choice, the daily work, lasting approximately 15 minutes. Afterwards, the second scientist repeated the activity for more than 15 minutes and then the third one did the same. Once the three scientists had been presented, the GAPP reporter introduced more specifically the gender issue providing food for thought with some questions about the advantages/disadvantages of being a female scientist, and the reasons why there are fewer women than men. An open discussion followed on the subject.

3.3.5   Tube Your Future (Nemo Science Centre, The Netherlands)

Tube Your Future focused on giving pupils a more complete image of the world of science and technology and offering an opportunity to meet professionals working in this field. By filming and interviewing professionals, pupils created their own image of the kind of work science and technology involves. Pupils were not offered a ready-made film. They asked questions following their own interests and prejudices. This gave them a chance to form a realistic idea about what kind of people work in science and technology.

The *Tube Your Future* Pilot Activity is a video contest that challenges pupils to interview and film professionals in the field of science and technology.

In order to carry out the activity, the NEMO science centre built a partnership of 5 members: Bètapartners (network of secondary schools, universities and organisations for the promotion of scientific studies), Jet-Net (Youth and Technology Network, a network between schools and technological companies), NIBG (The Dutch Institute for Image and Sound), Platform Bèta Techniek, VHTO (the national expert organisation on gender and science and technology).

The first step was the writing of the student and teacher manuals on how to carry out the activity started.

Later, the pupils participated in two *Tube Your Future* workshops. The first one 'Filming and Editing' was given by the NIBG (see above). Pupils had to make a complete television news programme, each group of five taking care of one item. For this they were provided with rough filming material to edit and a camera and microphone to film a self-made interview. The second workshop 'interviewing technique' was given in the classroom. The workshop consisted of several role plays in which the basic mistakes made in interviewing were highlighted.

Around halfway through the 3rd month each school was provided with a list of professionals that had previously given their availability to be interviewed and filmed, or had been previously contacted by NEMO. The organisation of the *final award gala* started.

The pupils started filming and editing in groups of five. Most of the groups were made up of boys or girls only, as requested by NEMO. The film, that had a maximum length of 4 minutes, could be uploaded on a special NEMO channel on YouTube.

Over the fifth month the glamorous award gala took place in the NEMO science centre. All participating pupils, teachers, professionals and partners were invited.

The gala started at 2 pm and the pupils and their teachers were welcomed by a red carpet and 'bubbley', lemonade for the pupils and *prosecco* for the teachers. The crowd was warmed up with a science quiz. Afterwards, the pupils followed a programme that included watching films by other pupils, short meetings with professionals and having a brief break during which they were allowed to walk around the science centre. All the films where shown in a random sequence order at NEMO's cinema.

Finally it was time for the award ceremony. Firstly the members of the jury were introduced. Secondly the eight nominees were announced. The makers of the nominated films had to get up and stand next to the stage, while a rap group of three participating pupils entertained the public. After this performance the jury started to give out the awards: The Golden Hammerhead for best technique, The Golden Lama for the most original film and The Golden Giraffe for the best film. Winning pupils posed happily for the camera with their awards and their newly-won iPods.

3.3.6   Movies and lessons (Youth Research Centre, University of Warsaw, Poland)

The aim of this activity was to strengthen the interest of young people, especially girls, in both the work and the life of scientists, and the cooperation with teachers having the purpose of broadening their knowledge on contemporary science, prospective possibilities of being a scientist. The activity unfolded with various phases: A training for scientists to raise awareness on the issue of scientific careers for girls; some meetings between teachers and scientists in their research centres; and a short movie showing experiments carried out by two young scientists, a man and a woman.

As part of the PA, the teachers were trained by the Youth Research Centre in the Institute of Applied Social Sciences at the University of Warsaw, The Copernicus Science Centre and the Partners Poland Foundation. During the visits in scientific centres the teachers could enter and meet scientists working at the Institute of Organic Chemistry and the Institute of Nuclear Physics of the Polish Academy of Science, as well as the Institute of Aviation.

Regarding the film, the plot for the script featuring young scientists talking about their scientific work and private life was prepared by experts. Scientific institutions were contacted and, through initial talks, the first choice among young scientists was made. 12 potential young candidates were chosen for the film, out of which 4 people were finally involved, two pairs of young scientists dealing with physics and chemistry. Then scenes were shot and the movie was edited in its final version. The movies were produced by professionals constantly cooperating with public television.

The invitations were sent out with the cooperation of the Education Department of the Warsaw Municipality to over 200 schools. On the first day, teachers were to be introduced to the GAPP background and took part in visits to research institutes during which they could ask their own questions to scientists about working in science and how it is to be a scientist. The programme of the second training day included a debate, presentations and workshops. The teachers discussed their own vision of science, their understanding of science, possibilities to work in science and the reasons for the difference in the interest in hard sciences between girls and boys. Then they watched the movies with young scientists, and set up some groups to get prepared for different tasks, exercises and topics applicable during the discussion on science. The teachers prepared individual lesson scenarios involving discussion with students, which were later evaluated by the team of experts and, after necessary changes, approved for implementation.

All of them followed a similar blueprint: Discussion and presentation of personal views on scientists, their professional and private life (frequently the aim was to produce some drawings, drama, performances etc.), presentation of the movie "Young scientists on themselves and on science", lesson/debate on the movie, stereotypes in the perception of scientists and their work, place of women and young people in science.

3.3.7 Evaluation of the Pilot Activities

Despite the variety in the Pilot Activities, there are some common points, both in the most successful elements and in the difficulties arisen during the implementation, which is worthwhile to highlight for the readers willing to experiment them.

The key element in the success of all the activities is the contact between the young people and the world of research. All the targets of the activities have maintained that the most positive element in their experience was the encounter with scientists and their workplace. This confirms the importance of the evidence achieved in the phases prior to the pilot activities implementation: The poor knowledge young people have about the identity of the people working in SET and on what they do.

Therefore, the reasons for this success and the change in the image of SET, as well as in the image of those working in the sector, are confirmed by the most recent studies on the world of the young people and science (M.C. Brandi et al. 2005 and C. Schreiner and S. Sjøberg 2004): The perception of science at school is very different from the one of the science in the "outer world". Also the first phase of the GAPP project confirmed that whereas science at school is seen as boring and "far away", other science communication forms are more positive and appealing. The "outer world" science, however, still is a quite unknown and usually regarded as a myth.

This is why, quite importantly, an effort has to be made to make those two images coincide, matching the two realities: During the Pilot Activities, the students asked for better contact to be made, when not provided, through practical exercises and experiments, aside from meeting scientists and researchers and listening to their accounts.

The idea underlying all of the six Pilot Activities is the contact between two worlds apart. A further element marking them is the participation level from boys and girls alike, not only as far as the attendance to the activities is concerned. A key point is how the six partners have built a contact between students and researchers, how they revealed the world of research to the students. The strategies ranged from an extremely active role played by the young people in activities where this active role of theirs emerges later on: From the *Tube your future* experience, which implied a total independence of the students in the responsibility of selecting, contacting and interviewing the researchers, to the training provided by the teachers and the projection of ready-made films for the students by the *Movies and lessons* project.

Concerning the focus on the gender, an evident result of the activities is the total gender equality in the participation of the young people, as demonstrated by their belief that they all have an equal potential. Moreover, the project has confirmed all the results from the statistics at European level, which reveal that girls are more inclined to study biomedical sciences, whereas boys are more interested in technologies. On the other hand, the interviews carried out during the evaluation of the Pilot Activities reveal that it is researchers that mostly express differences in the way men and women work; everybody agrees on the fact that potentially anyone can become a scientist.

As a practical recommendation, as the targeted public is made up of teachers, because they are the meeting point between the young people and the world of science, the Pilot Activities should represent an integration of their lessons and of the textbooks. Thus, a great deal of attention should be paid when implementing and linking the Pilot Activities to school curricula.

4   Discussion: A bottom-up approach. From the research on the public to the evaluation of the pilot activities

Across the countries involved in this research, the starting point is a perception of SET that confirms the current stereotypes, as shown in other large surveys such as the last Flash Eurobarometer on Young people and science (European Commission 2008).

In particular, the exact sciences are generally considered to be more difficult than human studies, as they would imply a special talent. On the other hand, technology involves a lot of physical grind: Being a hands-on activity, many students think that you need to be physically strong and that you get dirty, leading them to think that it is more a field for boys than for girls. Sciences and technology arouse spontaneous positive feelings amongst our participants. We have not found any conclusive differences between the two genders' knowledge of and opinions on education and job opportunities within the fields of science and technology.

It is worrying that students do not seem to have a clear concept of the professions available. Their frame of reference is primarily the subjects they have at school and the experiments they perform there.

In particular, participants call for a better knowledge of SET professions: All of our participants told us of the importance of practice in SET and see SET as strongly linked with reality.

In order to come closer to SET, gaining experience with science and technology since an early age has emerged to be very important. Moreover, a change in stereotypes has to be carried out starting from teachers, role models, from the meeting with SET professionals in schools, science centres, museums, and through the media, in fiction and advertising, by filming interviews with young and dynamic role models, both for boys and girls.

Practically, a direct participation and a "science and scientific careers in action" approach is to be desired as much as possible. Hence the launch of the Pilot Activities of the GAPP project: Under a *bottom-up philosophy*, they provide ideas for making young people closer to SET professions, at different school levels and overcoming the national differences among the European countries.

Acknowledgments

This paper was compiled by Federica Manzoli (Milan University), but the partners that have devised and implemented the qualitative research and the pilot activities in their countries are also co-authors. Thanks to Anne-Marie Bruyas of the Foundation IDIS-Città della Scienza of Naples, Carole Paleco and Olivier Retout of the Royal Belgian Institute of Natural Sciences, Paula Rabalo and Luís Barberio of the Pavilhão do Conhecimento-Ciência Viva, Amito Haarius, Leo Van den Bogaert and Marjolein van Breemen of the Science Centre Nemo, Marcin Sinczuch of the Youth Research Centre in the Institute of Applied Social Sciences at the University of Warsaw, Sheena Laursen and Lisa Klöcker of the Experimentarium Science Centre of Copenhagen.

References

Bauer, Martin W. & Gaskell, George. 2000. Qualitative researching with text, image and sound: a practical handbook. Sage: London
Brandi, Maria Carolina et al. 2005. Youth and Science in Italy: between enthusiasm and indifference, in: Jcom 4(2)
Cho, Seung-Ho et al. 2009. Images of women in STEM fields, in: Jcom 08(03)
European Commission. 2008. Flash EB n. 239, Young people and science. In http://ec.europa.eu/public_opinion/flash/fl_239_en.pdf [16.10.2009]

European Commission. 2007. Science Education Now: A renewed Pedagogy for the Future of Europe. In: http://ec.europa.eu/research/science-society/document_library/pdf_06/report-rocard-on-science-education_en.pdf [16.10.2009]
Howen, Harrison. 2008. Open Space Technology: A User's Guide, San Francisco, Berrett-Koehler
Gouthier, Daniele & Manzoli, Federica. 2008. Il solito Albert e la piccola Dolly. Springer Verlag: Milan
Gouthier, Daniele et al. 2008. The perception of science and scientists in the young public. Italian teenagers and science: views, beliefs and attitudes toward scientific research, in: *Proceedings of the IX PCST International Conference*
Manzoli, Federica et al. 2008. Children's perceptions of science and scientists. A case study based on drawings and story-telling, in: *Proceedings of the IX PCST International Conference*
Maxwell, J.A. 2005. Qualitative Research Design. An Interactive Approach. Sage Publications: London
Novotny, Helga. 2005. Unersättliche Neugier. Innovation in einer fragilen Zukunft. Kulturverlag Kadmos: Berlin. (it. ed. 2006. Curiosità insaziabile. L'innovazione in un futuro fragile, Torino: Codice Edizioni)
Schreiner, Camilla & Sjøberg, Svein. 2004. Sowing the Seeds of ROSE, Acta Didactica 4/2004, Oslo: Dept. of Teacher Education and School Development, University of Oslo, at http://www.ils.uio.no/english/rose/key-documents/key-docs/ad0404-sowing-rose.pdf (accessed 16 october 2009)

Munich's gender-sensitive education – extend girls' world to the fascinating field of engineering. Motivating girls to go for jobs with interesting tasks, career prospects and steady incomes

Barbara Roth

Abstract

Current economic, social and political conditions demand new strategies and innovative answers to challenges, such as demographic change or skilled labour shortage. In the competition for the most qualified and bright people women and minorities in particular have to be addressed. Although women and men nowadays have the same access to means of education, most of the women in Germany still choose a profession out of a list of ten, largely with low career prospects. Women remain a minority in natural sciences, technical and corresponding occupational study fields. The paper discusses the background of selected factors that may negatively influence girl's career choices. It will give an overview of the comprehensive gender initiatives and projects in all types of educational institutions in Munich since the nineties and describe the results. Successful projects in other countries and qualitative interviews with teachers and pedagogical staff indicate six main recommendations for motivating girls to choose jobs in STEM with better career prospects.

1 Introduction

The Munich city council acknowledges that equality and diversity are crucial for an innovative and dynamic city that offers a high quality of life to all its inhabitants, and promotes this conviction among its entire staff. Gender equality is an important aspect within this process. Its promotion represents an integral policy of the City hall and all its departments and has been furthered through the department of the equal opportunities commission since 1985. The strong legal involvement of the equal opportunities officer and her participation in important decision-making processes make her and her team a pivotal stakeholder within the "Gender and Diversity Movement in Munich". Schools and pre-school facilities develop projects and training relevant to gender equality.

In the field of STEM there are a lot of professions with very good career prospects, steady incomes and interesting tasks. In 2008 practically all engineers in Europe were employed. In July 2008, German companies were looking for 40,569 more engineers while only 4,013 engineers were looking for a job. A BITCOM[39] research reports 45,000 job vacancies in the ICT (Information and Communications Technology) market in September 2008. Ca. 60% of leading positions are filled with engineers (Bitcom 2009). According to meta studies on well-being and health the most important influence on health is self-efficacy and job satisfaction. According to the concept of "flow" (Csikszentmihalyi 2000) what makes you feel happy is the successful resolution of problems. As normally, engineers are hired for interesting tasks that include solving problems these activities may positively influence the wellbeing. According to basic human rights everyone has the right to choose a life based on individual talents, interests, and future prospects rather than on traditional concepts and to overcome role models.

2 Background: Selected factors that may negatively influence girls' career choices

Our conception of what women and men are and what they are supposed to be is produced by the society in which we live. Gender – the social „sex" of a person – is made by us in everyday lives in our interactions with others as a socially constructed category (www.genderkompetenz.info). This day-to-day, continuous production of gender has been called "doing gender" (Candance/Zimmermann 1987). Processes of "doing gender" are not only carried out in our society by individuals, but also through socially-standardizing practices such as legislation or the institutions of the family and marriage. "Doing gender" thrives on continually establishing a dual order. People are divided into two sex categories – boy or girl. From these categories, gender characteristics are derived, like preferring blue and pink, being technically or linguistically gifted and so on. In the course of her or his life the human being is "made" into a girl or woman or a boy or man in a complex process of rearing and education, social norms and values, stereotypes, identification, images, and traditions. Their influence on perception is mostly unconscious and thus very difficult to change. A wide range of influences may prevent girls in Germany from identifying themselves with STEM (Science, Technology, Engineering, Mathematics) or even as technically gifted in general.

Teachers and pedagogical staff have probably decided to pursue a career with relatively low income and low recognition for different reasons. One may be that they prefer interacting with people to any kind of technical occupation. In addition the pedagogical staff, especially in pre-school education, is mostly female with a traditional socialisation focused on nurturing and

---

39 BITCOM is the German Association for Information Technology, Telecommunications and New Media, representing 90 % of the German ICT market.

caring qualities. Nursery and primary school teachers in particular experienced no or very low technical training themselves and on average have only rudimentary training in learning by discovery and through technical experiments. Furthermore, most teachers have been instructed mainly in their special field. They acquire specialist knowledge in e. g. English, Latin or History, without any knowledge of the broad opportunities in the technical field. They have unconscious reservations towards engineering and technology. Without special training, teachers and pedagogical staff tend to transfer their own experience, attitudes, values and preferences to the children. Hence a gender-sensitive reflection on one's own attitudes and experiences should become a mandatory educational theme for every teacher and all pedagogical staff. In addition it is necessary that the proportion of male governesses, preschool teachers and elementary school teachers is increased.

On every list of famous German achievements there is much mention of engineering and technology products and services. The Federal Republic of Germany leads the world in many ways with regards to engineering excellence. Ever since the birth of industrialisation the nation of Germany has been an innovative force in these areas and many goods bear the 'made in Germany' logo. As a whole, products that are made in Germany have a good reputation of being well designed and of good quality. Traditionally German engineers perceive themselves as members of an exclusive closed inner-circle of sophisticated men. They assume "made in Germany" by German male engineers is a successful emblem. To give an impression of the attitude and culture in this inner-circle of male German engineers let me quote a German engineer in electro-technology and teacher at a vocational school in Munich. As I described what I have been working on and explained for him that schools in Munich are trying to motivate girls to study engineering, he answered: „I don't see the point in this. Actually German engineers are worldwide famous and very successful. That shows that the exclusion of women has proved of value". He was not joking. And it did not appear to him, that his comment was discriminating. His comment was full of conviction. As I learned later on, he has never been trained in the business advantages of diversified teams (European Commission Unit C4 2006) and Gender Mainstreaming and grew up in a male-dominated and male-educated environment. He never reflected upon what German businesses have lost because they have missed diversification in engineering completely. Besides these economical losses there is another very important aspect to this kind of attitude. In this atmosphere of a "male, white, middle class shark-pool" girls and women are at risk of isolation and exclusion in technical courses in schools, in apprenticeships, at university and at work. As a result all the staff – male and female – have to be trained in Gender Mainstreaming.

Especially in IT the stereotype of the geeky-programmer, pale, dark circles around the eyes, badly dressed with a weird haircut, stuck behind his computer for at least 16 hours a day and not capable of speaking in whole sentences, acts as a deterrent for girls. Most of the girls want to be attractive, hence IT should become less geek and more chic!

A transgender IT specialist, managing director of a fast growing web hosting solution provider company, who was male for 29 years, reported after her gender transition, that she had to face problems she (he) did not have before becoming an attractive very feminine woman. Since transitioning she reports that it was really difficult to get guys to take her seriously. (Whitehead 2008). It was very difficult to establish credibility she had been granted in former days when she used to be a man. "She is concerned that IT is not seen as a sexy career path: There is a view that if you want to get on in IT, you can't be feminine, you can't be attractive."(Armstrong 2008).

Common general prejudices in Germany influence girls' career choices. The strong traditional picture of the German housewife, "who is in charge of saving the Universe" puts heavy pressure on women who go for a career for themselves. According to German tradition, women are meant to be the upholder of moral standards, responsible for development of social life, volunteering, and raising children. These kinds of prejudices appear in publications regularly, even in famous and acknowledged German newspapers. Two prejudices in particular seem to be a huge burden to women. Prejudice number one is that children are only happy and develop positively if they have been raised by their mothers alone. Children need "the heartbeat of their mother" the whole day long (Meves 2007). Full-time day nurseries, kindergartens and schools lead to mental disorders and developmental problems of children, such as drug addiction, hyperactivity, learning disabilities, and lack of concentration. Prejudice number two is that young women who judge a mixture of professional work and child care as ideal are a minority. If mothers have the choice between staying at home with their children or going to work following a professional career most of them stay at home and take care of their children because caring and nursing is the real nature of woman (Gilbert 2009). Both prejudices are thoroughly proven as wrong, though highly recognised in German society.

In addition one of the strong factors that may negatively influence girls' career choices is television. In Germany on television nearly 50% of the professions represented are limited to three occupational areas. These are public order and security, media and humanities/arts and TV-specific professions. This is followed by health (7.6%) sports (6.1 %), education and social affairs (4.6%) metalworking industry (0.7%) science (0.4%) and computer science (0.3%) the latter including related professions (Mangold 2008). Due to this under-representation of STEM in Media there is a lack of role models in the media for (female) engineers.

3   Educational system in Munich and the role of gender and diversity

In Germany, each of the 16 federal states is responsible for its educational system. Furthermore, so-called "free cities", apart from the capitals, exist in most of the federal states. These cities are allowed to establish their own education concepts within the legal framework of each federal state. Thus, the city of Munich has a traditional right to establish its own schools, kindergartens and crèches (day nurseries) and is Germany's greatest provider of educational and learning services, as far as German municipal councils are concerned. It is mandatory for every child from the age of six to attend school for twelve years. Besides, every child has the right to attend a kindergarten (from three to six years). From 2013, this right will be extended to include children who are under three years old.

Several educational facilities In Munich prepare for university. All children attend the compulsory primary school (Grundschule) until grade 5. Depending on their success in primary school they switch to a Gymnasium or a Realschule or a Mittelschule afterwards. The latter is not listed here, because it is not directly preparing for University. The names of the educational institutions in Munich are: Kinderkrippe (Crèche or day nurseries, under three years old), Kindergarten (three to six years), Grundschule (Primary School, six to ten years), Gymnasium (highly academic Secondary School qualifying for universal university entrance), Realschule (Secondary School, preparing for vocational training or post secondary schools), Fachoberschule and Berufsoberschule (Post Secondary Schools), Meisterschulen and Technikerschulen[40] (advanced vocational qualification, Schools for master craftspeople). 40% of the students at universities in Bavaria acquired the university entrance qualification at a school other than a Gymnasium[41].

Essential to successfully motivating girls for STEM is the right attitude, conviction, and knowledge of staff working in (pre-) school settings. Hence the city of Munich trains the staff on aspects of gender diversity and gender mainstreaming in different circumstances, and the consultant for gender-sensitive education for Munich schools has developed multiple concepts of education and training for teachers. Every school has to assign a teacher the position of girls officer. She is trained regularly, supervises the implementation of gender mainstreaming in daily school life and organises special projects. Furthermore, a network of educators who organise events, such as the "Girls' Day", meet regularly for professional development. It is mandatory

---

40   Master craftspeople and comparable qualified people are awarded a Higher Education Entrance Qualification since October 2008 for Universities of applied science.
41   The Chamber of Crafts for Munich and Upper Bavaria awards the title „Master of Craftsman". The Chamber represents about 67,000 member enterprises with about 280,000 employees.

for Headmasters to attend training sessions on gender mainstreaming. Some schools assigned a teacher as boys officer and these meet regularly as well. Gender equality is also a topic for kindergartens. Since 1998 a consultant for gender-appropriate pedagogy in Munich Kindergartens has been training staff in gender mainstreaming and gender sensitive education in various ways. The aim is to integrate the internal and external world of the children in the pedagogical work and enhance the individual strengths and abilities of each child as a counter balance to social pressure and traditional role expectations. This includes regular meetings of staff to review their work, materials and habits (e.g. books, "dress-code" of the children, costume-materials, access to technical materials, motion and body exercise, etc.).

The main four prerequisites for staff working in (pre-) school settings for successfully motivating girls for STEM (Budde 2008; Roth 2009) are the following.

Firstly, they must be able to recognise that boys and girls may be being differently treated in their establishments; thereby recognising that some adverse socialisation may have occurred and that this needs a counter balance.[42] Secondly they need to ensure that boys and girls have equal access to activities of technical nature. This may mean allowing only girls to use certain materials at certain times. Thirdly, they have to ensure a gender-sensitive learning environment for both sexes. Fourthly, they may need to give explicit help to girls to encourage their interest in technical activities.

4 Description of some projects to motivate girls for STEM

The desired goal of education of the City council of Munich is to make participation in all aspects of life possible, provide children with opportunities and enhance gender justice. The following selected initiatives and projects for children, adolescents and young adults successfully motivated girls for STEM or at minimum improved their competences for a career in STEM.

4.1   The Bavarian Education Curriculum for Kindergartens (BEP, "three to six years old") on Nature, Science and Technology and the project "Tiny Tots Science Corner" (Haus der kleinen Forscher/innen 2009)

The BEP and the Tiny Tots Science Corner are meant to give the children the opportunity to experience nature and technology and strengthen their natural

---

42 One Kindergarten head master with a very successful technical education calls her Kindergarten "a rubber boots Kindergarten" („Gummi-Stiefel-Kindergarten") meaning every girl (and boy) has to wear clothes suitable for any kind of movement at any weather.

curiosity and guide children's eagerness to experiment by giving them a playful introduction to the natural sciences.

The Tiny Tots Science Corner initiative consists of various basic building blocks including workshops, materials, an internet platform, a certification of Kindergartens as "Tiny Tots Science corner" and a national day of action called "Early Discovery Day" to encourage Kindergarten teachers to integrate more natural science and technical content into their programming. The BEP is a compulsory curriculum for the Kindergartens and includes technical projects. The accompanying publications on BEP offer a wide range of examples.

A one-day introductory workshop on Tiny Tots Science Corner provides educators with basic pedagogical knowledge and prepares them to carry out the first experiments e.g. with "water". The half-day follow-up workshop gives participants the opportunity to share their experiences.

The Tiny Tots Science Corner will provide participating Kindergarten facilities with experiment description cards grouped into "projects" focusing on individual topics, such as "water," "air," "light and colour," etc. Each project consists of approximately ten experiments. These experiment cards are reported as very helpful, because staff in pre-school facilities tend to be inexperienced in technical matters and feel encouraged to use the material via the simple and short explanations on the experiment cards.

The Tiny Tots Science Corner pilot program has been running in Berlin since 2006. In October 2007 three delegates from Munich were trained in the concept and the material of the Tiny Tots Science Corner. From January 2008 to July 2009, 292 of the pedagogical staff were instructed in Munich. The pedagogical counsellors expect about 80% of the participants to work with the materials in the Kindergartens and day-care facilities (Roth 2009). Are the BEP and the Tiny Tots Science Corner gender sensitive STEM projects? The educator supervisors for Munich trained by the project team of Tiny Tots Science Corner in Berlin did not report a gender sensitive aspect in the training. The German title of the project excludes female children: "Das Haus der kleinen Forscher" – in German there is both a she- and he-form for job descriptions, which is ignored. The German title of this project only mentions the masculine form for "Forscher" (male researcher/scientist) and addresses only the boys. Because the council of the City of Munich requested it, there is a (in the internet unreadable) subtitle "Naturwissenschaft und Technik für Mädchen und Jungen" (natural sciences and technology for girls and boys) and is printed in very small letters on every paper publication. The web page of "Tiny Tots" does not refer to any gender sensitive aspects. If one, like in an article on the web page of "Tiny Tots Science Corner", addresses only female educators in Kindergartens in Germany that almost represents the reality in Munich with only 1.5% of Kindergarten educators being male. It may help promote the cause of increasing the proportion of male educators in Kindergartens if both male and female educators were addressed in every kind of publication. The BEP is a curriculum and refers to gender questions in a special clause about gender-aware education in general but not specifically related to natural sciences, mathematics and technical instruction. The additional publication on the BEP (Bayerisches Staatsministerium für Arbeit und Sozialordnung 2005) gives in the section nature and science general advice on the realisation of different projects but does not refer to gender topics there. The gender sensitive aspect of technical education for the "Tiny Tots Project" and the BEP in the STEM field is left alone to the Kindergarten educators. The department for Kindergarten education of the City of Munich trains the Kindergarten educators on the gender – sensitive aspect via the educator supervisors, the so called regional quality managers for STEM in Kindergartens and the consultant for gender justice in pedagogics for Kindergartens of Munich. They offer a range of education and training in the field of gender equality and gender-sensitive teaching for the staff.

4.2   Toy-free-time in crèches or day nurseries and kindergartens

The project toy-free kindergarten was developed in the district of Upper Bavaria in Germany in 1992 by members of a study group initially in order to prevent addiction. One of the basic considerations was that the competences of children may be effectively enhanced if the children themselves become the creators of learning processes. The aim was to recreate scope for playing as well as fantasy and creativity and thus enhance self-affirmation and self-assurance (aJ Aktion Jugendschutz 2009).

The results of the accompanying study showed that the following competences are especially enhanced by the "toy-free kindergarten" project: Creativity, autonomy, capacity for play, problem-solving, critical thinking, technical interest, and psychosocial competence (Winner 1996).

Normally the following basic rules and features of "toy-free-time" are applied. For three months all toys and materials for handicrafts (such as crayons, paper, scissors, tools) are removed from the rooms of the kindergarten. Only the furniture, blankets and pillows remain. At the children's request and if there is no other possibility, required materials and tools can be used, but no toys. The kindergarten teachers commit themselves to observing the children actively and to be present as partners. However, they do not make any offers, do not provide any rash solutions and do not place any substitute materials at the children's disposal. The concept of toy-free kindergarten needs detailed planning and preparation of the project by a professional team. For the first project in a kindergarten experts accompany the project and advise the educators. The first project started 1992, it was evaluated and the documentation was published 1996. Over the years after a lot of practical experience with the toy-free-project, different kindergartens developed their own style in realising the principles of toy-free-time in kindergartens (Roth 2009). Different kindergartens vary the toy-free-time project in time, strictness, observation, and evaluation. The educators commit themselves to the toy-free-project because they are convinced of the special opportunities this concept includes. Every kind of toy-free time in kindergartens requires good cooperation with the parents, grandparents and all psychological parents of the children.

I would like to point out that any and all toys are removed from the play room, and even materials such as crayons, paper, scissors, tools, etc. are no longer available. Only the furniture is left in the room. The children are informed about the toy-free period in detailed conversations. While some groups put their toys into an adjacent room together with their educators, others found their rooms empty at the beginning of the project. At the children's request and after mutual consent, required materials and tools can be used, but no toys. Thus the children have to come up with their own ideas about what they want to do. They have to decide, communicate their decision and develop a technical spirit of research, while working with the simple crafts materials they had requested and were offered. In addition to the psychological aspects they train their manual skills as well. In general the project toy-free time in Kindergartens especially fosters imagination, creativity, communication competence and self-confidence – all prerequisites for a successful future in engineering.

Kindergarten educators, who had been trained in gender-sensitive aspects, reported brief interventions necessary because of quarrels over the tools. They had to intervene in order to make the tools available for the girls as well. Some reported the necessity to encourage girls in a very strong way in the beginning. After a while the boys and girls interacted equally enjoying the scientific and constructive play. As a result the *„[c]ontact with the other sex has become simpler. Prior to the project they used to play almost exclusively with playmates of the same sex, now this is not so important."* (Schubert/Strick 1994: 17) In addition to all the advantages named earlier, the project encourages girls to enjoy technical manual work and to interact with the boys in this field. It strengthens the self-efficacy and self-esteem of girls. It is a project that integrates the natural curiosity of girls in the day to day experience in day-care facilities and encourages them to keep up their technical interest even after the toy-free time.

4.3   "Girls' Day"

Girls' Day is a nationwide career orientation day. Numerous companies, businesses, research labs and institutions invite female pupils in grade 5 through 10 (age 10 to 15) to visit their places of work for one day, opening up their labs, offices, workshops and editorial offices. Girls' Day provides female pupils with an opportunity to gain practical insights into the world of work, particularly in technical and technology-related fields. The focus is on hands-on practical experience.

One of the goals behind this nationwide activity Girls' Day is that female pupils see how interesting and exciting work such as that of IT specialists, biophysicists or car mechanics (repairwomen) can be. Girls test their skills in hands-on activities and experiments, allowing them to widen their career spectrum. Previous impressions can be reinforced while misinformed views about the world of work can be corrected. Becoming familiar with new unknown careers opens up girls' eyes to their own career options. It also gives them a chance to make contact with potential employers early on. Beyond that, decision-makers in companies and universities become aware of the diverse strengths and skills of today's female pupils.

Every year the Munich superintendent of schools, Mrs. Weiß-Söllner, recommends in a personal letter to all head masters to send girls to the Girls' Day. Included in this letter is more information on the Girls' Day. Furthermore the teachers who are responsible for the organisation in the schools are invited to the Pedagogical Institute for an information afternoon about the Girls' Day and an accompanying contact person at the Pedagogical Institute is named. In April 2009 in Munich 27 municipal schools took part in the Girls' Day, most of them with several classes (each class containing 25 to 32 pupils). That is to say that at least 1200 girls of municipal schools attended the Girls' Day 2009. Currently parallel to the Girls' Day a Boys' Future Day is being established. In 2009, 14 secondary schools sent boys to a Boys' Future Day encouraging boys to consider jobs where the male sex is underrepresented such as hospital nurses or primary school teachers and Kindergarten educators.

In 2008 a reliable and significant verification on the effects of the Girls' Day did not yet exist. The results of a small enquiry showed that, 52% of the girls who participated in the then so-called "girls and technology day" in 1990 in Munich decided to go for a career in STEM (Ihsen 2007).

4.4   Special lessons in "self-confidence training" for girls

The aim is to enhance the self-esteem of the girls so far that they are sufficiently self-assured to develop their personal abilities and make choices without limitations set by strict gender role expectations, thus widening the array of possible career choices for girls to the field of STEM.

A wide range of schools in Munich offer "self-confidence training" for girls. The Pedagogical Institute of Munich has been training female teachers in "self-assertion for girls" for many years. When the teachers have achieved a formal specialist qualification in this they regularly give courses for girls in their schools in the afternoon.

4.5   Special lessons Information Technology (IT) for girls

The goal of this initiative is this, to motivate girls for IT. Thus, a couple of secondary schools (Realschulen) and some Gymnasiums (Grammar schools) in Munich offer special Information Technology lessons for girls. In some schools the regular Information Technology lessons are separated into girls and boys only groups. Other schools offer special courses for girls in the afternoon. In reality, what is in fact taught and in which manner depends largely on the teachers. So the encouraging effects of girls-only lessons in Information Technology seem to be influenced by the content of the lessons and whether the teachers have been trained in gender-sensitive education. According to a survey by the British Computer Society, motivation for girls to go for a career in engineering or ICT (Informatics and Communications Technology) seems to depend on the content of the courses. Some girls said that they did not want a career in IT because "it is just being a secretary" (FT 2008). Further research showed that these girls had attended ICT lessons in schools which taught them only how to handle text – processing software or how to improve their touch typing abilities. But girls prefer creativity and problem-solving tasks and want to be challenged. So they decided not to go for a career in ICT because they believed typing is ICT („Would you enjoy a cookery course, if you only learned opening cans?") They had never learned anything about the reality in ICT careers.

4.6   Girls-IT-clubs

Girls should be encouraged to develop web pages, should be instructed in programming and equipped with hardware knowledge, thus be empowered to enjoy the creativity of information technology and the "IT-Spirit". A couple of schools organise Girls' IT clubs, some in Cooperation with "KommIT – Frauen in IT- und Multimedia", "Women in ICT and Multimedia". The teachers encourage girls to build web pages, enhance their Internet research abilities and become skilled mechanics in regular lessons in the afternoon.

4.7   Differentiated co-education[43] in physics at the St. Anna Gymnasium

Until 1987 the St. Anna Gymnasium used to be a girls (only) school with focus on mathematics and sciences. During these years the same number of girls took up so-called Leistungskurse, that is, advanced courses, in mathe-

---

43   Differentiated co-education means, that special lessons are taught in "girls only" and "boys only" groups, that is to say mono-education or mono-educative teaching

matics, physics or chemistry as boys in co-educational schools. At that time the results of the girls used to be very good in these fields. After the school became co-educational and more and more boys attended the school, the proportion of girls in Nature and Technology advanced courses dropped to the national average, e.g. 11% in Physics in 1995. The head of the physics department, Verena Schroll, and her female colleagues observed that they themselves had all attended "girls-only schools". They therefore started a so-called differentiated co-education project.

The idea was to develop girls' self-confidence within their own capacity in physics in a girls-only group in physics (so called mono-education lessons) so that they would be able to compete with boys in the following years. Meanwhile they would continue to have educational advantages of a co-education system, girls and boys together for other lessons.

In grade 8 (age 13), girls-only groups were formed for physics lessons. A team of female physics teachers developed physics lessons that took girls' interests into consideration, for example more medical or biological contents, and exchanged ideas with each other about their new materials. The results proved that Verena Schroll and her colleagues' assumptions were right. Even only one year, "girls-only" physics lessons in grade 8 combined with teachers' training in gender-sensitive education, greatly improved the girls' motivation for physics. The school achieved a sustainable proportion of 25 to 40 percent female pupils in advanced Physic courses in grade 12 and 13. That is more than twice the average in Bavaria (Schätz 2002; Schroll 2004).

4.8 Differentiated co-education in mathematics at the Luisengymnasium

The assumption of Ulrike Schätz, a mathematics teacher at the Luisengymnasium, was that if girls were taught mathematics in a mono-educational situation they would achieve better results. In grade 11 the girls were offered the possibility to decide if they wanted to attend a mono-educational or a co-educational course in mathematics in grade 12 and 13.

An impressive study showed that the girls who had chosen mono-educational instruction were much more successful in mathematics than those girls who were instructed in co-educational courses. Further research proved that the girls who had decided to attend the mono-educational mathematics course had been less successful in grades 8 through 11, than the other girls of the classes. Both groups of the girls had to write the same final exams, the central state Bavarian Abitur[44], in mathematics. The results of the girls of the mono-educational mathematics courses were on average one mark better than

---

[44] Diploma from German secondary schools qualifying the student for university admission or matriculation

the results of the reference group on a scale of 6 possible marks. The observations of teachers at the St.-Anna-Gymnasium and the Luisen-Gymnasium were strongly confirmed by a study of the institute for pedagogical psychology of the Ludwig-Maximilians-University of Munich in 1995. About 1300 girls were tested, most of them in mixed classes. The reference classes of the St.-Anna-Gymnasium were girls-only classes. Most of the teachers in mixed classes had even attended reattribution training. They had learned a method of verbal interaction with girls in lessons in order to substitute negative preconceptions with a positive present situation. The girls instructed in co-educational or mono-educational lessons were surveyed (for a comparison) according to motivation, interests, and attitudes towards physics. The results were amazing. Confidence in their own capacities was the same before their first physics lesson in all the groups of girls. After three months, the survey showed a clear growth in confidence of the girls in mono-educative girls' classes, while a decrease in confidence of the girls in mixed classes was observed. Interest in physics was significantly influenced positively by girls-only lessons. The girls accepted their unique situation in sex-separated courses in a co-educative school more and more. Their attitudes towards physics showed a distinct difference. Girls from mono-educational classes had a higher motivation for physics during leisure time, for active participation in lessons, for choosing physics courses when possible and for taking up a profession including physics.

A more recent study (Kessels 2002) confirms the former results and proves that a temporary separation of puberty-aged boys and girls in physics increases the number of girls motivated for this subject. As a consequence, girls do not exclude training for a profession in physical or STEM- fields from the start. Especially during puberty, a time of searching for individual and sex-roles, girls hardly identify with a subject as strongly male-connoted as physics. During mono-educative instruction one can obviously find a situation where sex becomes an irrelevant category.

4.9 Differentiated co-education in a couple of secondary schools in Munich

The aim is to give a counter balance to socialisation influences to both sexes, girls and boys. Several secondary schools (Realschulen) in Munich have developed a concept of gender-sensitive separated courses for girls and boys for different subjects, like mono-educational lessons in physics for girls and boys-only courses in languages. In addition "sexual education", "HIV-prevention", "self-confidence training", "stop smoking", "talent-development", "IT-Clubs" and "career choice" education lessons are taught in girls and boys only groups. The results are currently in discussion.

4.10 A school with a career choice project – the Willi-Graf-Gymnasium

The career choice project at the Willi-Graf-Gymnasium targets at broadening the horizon of girls for less common professions. The girls' officer and counsellor for grades 5 to 8 at the Willi-Graf-Gymnasium, Mareile Müller, combines different projects in grade 8 in order to motivate girls for STEM. The girls start with a project (5 hours) "Princess, nurse or mother?", developed and realised by an independent association called MIRA. The aim is to work with the girls on the opportunities and limitations of a woman planning her life and to point out the multiple choices they have. The project includes the following objectives: The girls appreciate their competences and abilities. They take a positive attitude towards themselves. They verbalise their hopes and desires for their future. They realise different options. They identify the first steps towards their future.

Mareile Müller reported that the educators were struck by how unrealistic and low pitched the expectations of the girls were. This first project opened the minds of the girls toward the option of a future in STEM. The following two days the girls took part in different STEM related projects organised by "girls do technology" ("Mädchen machen Technik"), from the Technical University Munich. The girls worked in a strongly self-developing and experimental way in different groups in the fields of robotics, chemistry, architecture and biology. The girls successfully constructed, programmed or analysed.

The feedback from the girls was very positive. They praised the creativity, the team-play, the self-activeness and the discovery-oriented work atmosphere. After these three days, a lot of them mentioned that they could imagine working in the field of STEM:

4.11 "School Meets Science" – the municipal girls-only Bertold-Brecht-Gymnasium (BBG)

"School Meets Science" wants to inspire girls for a career in research in natural sciences. The BBG is the first public girls-only school in Bavaria with a natural sciences and technology focus and an ongoing co-operation with the "Helmholtz Zentrum München – German Research Center for Environmental Health". The focus on nature and technology means that the girls occupy themselves more with natural sciences (chemistry, physics) and informatics. Day to day lessons are enriched by real research examples of the Helmholtz Zentrum München, e.g. applied statistics in Mathematics via an example of heart attack or stroke statistics gathered by scientists from the Helmholtz Zentrum München. In Addition, several lessons concentrate on experimental work in small groups. Scientists and teachers cooperate on special projects. Furthermore The Helmholtz Zentrum München invites classes from grade 5,

7, and 9 (age 10 to 14) to an interdisciplinary science project week, where they also meet female role models. In addition to the practical and experimental experience the girls gain presentation-experience and receive positive feedback from the visitors, which is a huge motivation for the girls.

The project "School Meets Science" is evaluated with support from the IPN of the city Kiel. The study focuses on the image of natural sciences in the evolving pupil. It is designed as a long-term study that goes along with the project. First results show that the understanding about the general functioning of general sciences is positively influenced by the project.

5 Interviews, analyses and reports of teachers and pedagogical staff[45] (Roth 2009)

The pedagogical staff mentioned the following hindrances and support for their work. As main hindrances they described having not enough room (e.g. Kindergartens have no extra room for experimental learning, and not enough room for storing the experimental material), not enough money for experiments, technical equipment and computers, classes are too large (up to 32 pupils per class) for experimental teaching or self-discovery individual learning, not enough teachers, not enough time for preparation for the teachers (e. g. at Rosseau Lake College in Ontario, Canada teachers give 16 lessons per week, in Bavaria teachers at Gymnasiums give 24 lessons per week), not enough room for teachers in schools to work and prepare lessons and projects in cooperation with colleagues (in many schools teachers do not even have their own desk), "when special educators leave the project dies" – typically a lot depends on single persons, over-loaded curricula in nearly every field and hence not enough time for example- and discovery-oriented learning. As supporters they stated very committed and highly motivated individual teachers and pedagogical staff, who work together in the evening and at weekends; pedagogical freedom; support from the headmasters; support from the municipal council; support from the colleagues or an interacting team of colleagues; associations which support teachers, like MIRA or KommIT; Co-operation partners like the Helmholtz Zentrum München or the TU Munich (Technical University Munich); support from businesses, e.g. the group of companies that finance and accompany the project "tiny tot centre", and last but not least gender-sensitive colleagues.

The pedagogical staff put some prerequisites for staff successfully motivating girls for STEM. Teachers and pre-school educators should be open to the fascinating world of science and engineering and sensitive for gender related obstacles and supporters for learning. Their behaviour should be

---

45 On the basis of these results a pilot project will be developed by the Pedagogical Institute Munich

based on a reflection on their owns gender stereotypes and gender-role (e.g. "teachers used to help girls and explain to boys"). They should be innovative, allowing girls to develop a research spirit in experimental self-learning.

Continued education and training for teachers and educators should include mandatory training in gender mainstreaming with practical observation exercises and role plays for educators and teachers; improving skills of observation and perception of gender related behaviour and equipping educators and teachers with a wide range of adequate reaction concepts as a counter balance to socialisation influences and a wide range of experimental projects.

Additional postulations are: Kindergartens and Crèches need an extra room for experimental learning; educational facilities need more money for experiments and technical equipment; experimental or self-discovering individual learning requires smaller classes; teachers need more time for preparation in schools and fewer lessons per week and more room in schools to work and prepare lessons and projects is needed. An indispensable prerequisite for experimental and self-discovering individual learning is an over-view and unload of curricula in nearly every field. The new curricula should be focused less on content and more on example-oriented life-long-learning methods. Sustaining the natural curiosity and interest of children in experimental technical learning has to begin when they enter any kind of educational facility. Reliable gender quotas for studies and vocational training have to be implemented: At least 30% male day care assistants, 30% male Kindergarten teachers and 30% male primary school teachers. Most of the teachers argued, that differentiated co-education with girls- only groups in Mathematics, Physics, Nature and Sciences can become a benchmark for quality in educational facilities.

6 Recommendations and outlook

Most of the described projects are examples out of the daily -life in (pre-) schools. Practical experience and qualitative interviews are often the most successful way to find out, what is in fact helpful for achieving the goals, proving whether assumptions are valid or receiving precious hints for developing new strategies. In this paper the most significant conclusions are:

Firstly, a lot of evidence points out clearly, that mono-educative "girls-only" lessons or schools motivate girls for STEM with the highest results up to now. This seems to be an international phenomenon, with reporting countries including Australia, France, and Austria.

Secondly, a simple and basic pedagogical principle is that before preparing lessons, a teacher has to consider in depth the individual precondition of every pupil involved. Considering gender means thinking thoroughly about all socialisation influences that might have occurred. This requires a high-level of gender-knowledge in addition to self-reflection and awareness of the own sex-role of each teacher and educator in (pre-) school. Without this a gender-sensitive reflection on teaching is not successful. Therefore every female and male teacher and educator in (pre-)school facilities has to be trained in gender-sensitive aspects of learning and teaching.

Thirdly, technical education has to start very early, before socialisation influences "kill" girls' natural curiosity about nature and science.

Fourthly, the method and approach used in teaching natural sciences or any STEM-related subject determine the success. Gender–sensitive teaching aiming at motivating girls for STEM can counteract socialisation and should be inquiring, experiment orientated, including problem-solving and practical examples out of the daily life of the girls. It should use the girls' creativity and be challenging.

Fifthly, we are in need of mandatory and reliable quotas for male trainees in studies and vocational training.

Sixthly, diversity initiatives should be linked to accountability systems and tools to measure progress. Educational facilities have to do what you usually do in business: Set goals, plan, check, and act if results do not comply with the scheduled goals.

Even though policies regarding diversity and gender-equality for every kind of educational institution have already been established and enforced for many years the situation in STEM for women has worsened in Germany. The number of female students in engineering and especially female participants in ICT (European Commission „She-figures 2006") has been decreasing for many years. There are already a wide range of excellent and thoroughly investigated examples of good practice and a huge number of concepts for increasing the number of girls and women in STEM. However, we have to reconsider the implementation of such strategies and manage and monitor the process accurately for change (Harvard Business Review 2004; Thomas 2009; McKinsey&Company 2008). Schools need a new quality and innovation management (Cobb 2003). These process skills are crucial for future progress. Thus, diversity initiatives should be linked to accountability systems and tools to measure progress. In other words, schools, kindergartens, crèches and pedagogical institutes have to learn to act like companies. They have to do what you usually do in business: Set goals, plan, check and act if results do not comply with the scheduled goals.

References

aJ Aktion Jugendschutz Landesarbeitsstelle Bayern e.V. 2009. toy-free Kindergarten, www.spielzeugfreierkindergarten.de, 22 Februar 2009
Armstrong, Lindsay. 2008. Less 'geek' more chic is the way forward, in: Financial Times, U.K., Special Report Digital Business, May 14: 2

Bayerisches Staatsministerium für Arbeit und Sozialordnung, Familie und Frauen & Staatsinstitut für Frühpädagogik. 2005. Der Bayerische Bildungs- und Erziehungsplan für Kinder in Tageseinrichtungen bis zur Einschulung. 2., aktualisierte und erweiterte Auflage. Weinheim, Beltz: Basel

BITCOM. 2009. http://www.bitkom.org/de/themen_gremien/54621.aspx (1 March)

Budde, J.; Scholand, B.; Faulstich-Wieland, H. 2008. Geschlechtergerechtigkeit in der Schule, Juventa: Weinheim

Cobb, Charles G. 2003. From quality to business excellence: a systems approach to management, Milwaukee, Wisc.

Csikszentmihalyi, Mihaly. 2000. Das Flow-Erlebnis. Klett-Cotta: Stuttgart

Desvaux, G. & Devillard, S. 2008. Women Matter – female leadership, a competitive edge for the future, McKinsey&Company

European Commission. 2006. Community research Unit C4: women in science and technology – the business perspective

European Commission community research. 2006. „She-figures 2006" http://www.earlytechnicaleducation.org/chapter2.html (1 March 2009)

Financial Times UK. 2006. Special Report Digital Business: Where are all the women in IT? November 8

Gilbert, Neil. 2009. Interview with the soziologist in the German Newspaper „die Zeit" 26 Februar, http://www.zeit.de/2009/10/Interview-Gilbert (8 March 2009)

Harvard Business Review. 2004. Vol. 82, No. 9, September „Diversity as Strategy"

Ihsen, Susanne. 2007. Technische Universität München: How does the GirlsDay change engineering education in Universities, Documentation of the national symposium „Girls'Day" Berlin, 5 December

Haus der kleinen Forscher. 2009. Naturwissenschaft und Technik für Mädchen und Jungen; http://www.haus-der-kleinen-forscher.de, 28 February "Tiny Tots Science Corner"

Kessels, Ursula. 2002. Undoing gender. Eine empirische Studie über Koedukation und Geschlechtsidentität im Physikunterricht, Weinheim/München

Lafortune, Louise. 2008. Un modele d'accompagnement proffessionnel d'un changement, Quebec

Mangold, Michael. 2008. Die Berufswelt im Fernsehen, in: Macht.Fernsehen.Frauen, Einfluß des Fernsehens auf die Berufsorientierung von Mädchen und jungen Frauen, Bündnis 90/ Die Grünen NRW (Hrsg.)

Margolis, Jane & Fisher, Allan. 2003. Unlocking the clubhouse: women in computing, MIT Press

Meves, Christa. 2007. Die ersten Jahre sind grundlegend, Interview printed in the FAZ, 06 Februar

Roth, Barbara. 2009. Multiple personal Interviews with Kindergarten headmasters, educators, teachers and quality managers for gender justice in schools, not publised, contact: barbara.roth@muenchen.de, Padagogical Institut Munich Germany

Schätz, Ulrike. 2002. In Wortprotokoll des Bayerischen Landtags, 14. Wahlperiode, 6. Februar 2002, Stellungnahme der Experten, Anlage 4: 71ff.

Schätz, Ulrike. 2001. In: Bramkamp, K. and Jost, M: Hochbegabte Mädchen und Frauen Begabungsentwicklung und Geschlechterunterschiede higly gifted girls and women, gender differences and talent development in Labyrinth, DGhK, 69: 18ff.

Schubert, Elke; Strick, Rainer. 1994. Toy-free Kindergarten, aJ Aktion Jugendschutz Landesarbeitsstelle Bayern e.V.
Seiter, Josef (Red.). 2007. Technik – weiblich, Analysen zu mädchen- und frauenzentrierten Fördermaßnahmen im Bereich von Technik und Naturwissenschaft; Schulheft 32. Jahrgang, Studienverlag: Wien
Schroll, Verena. 2004. Monoedukativer Physikunterricht – Gute Noten für Mädchen, www.leaNet.de (17 Februar 2004)
Thomas, D. 2009. „IBM finds profit in divesity" http://hbswk.hbs.edu/item/4389.html (15 March 2009)
West, Candance & Zimmermann, Don H. 1987. Doing Gender. In: Gender and Society 1 (2): 125-151
Whitehead, Peter. 2008. No longer a member of any boys' club, in: Financial Times, U.K., Special Report Digital Business, May 14: 6
Winner, Anna. 1996. accompanying study toy-free kindergarten, aJ Aktion Jugendschutz, Landesarbeitsstelle Bayern e.V.
www.genderkompetenz.info; Humboldt University Berlin [2009.02.18)

Mentoring and video clips – ways against stereotype threat

Sylvia Neuhäuser-Metternich and Sybille Krummacher

Abstract

Two different but related approaches to change the image of science and scientists and thus reduce stereotype threat for women in science, engineering and technology (SET) are presented and discussed here: 1. Mentoring is a support concept where through the interaction with their mentees, educators and superiors as mentors are made aware of the discouraging forces experienced by female students. 2. Video clips about female engineering students accompanied by information material to attract girls towards science or engineering studies were produced at Dortmund University of Applied Sciences and Arts. Both approaches are aimed at changing the perception of SET and of girls' and young women's competences in this area as well as the attitudes of peers and educators at school and universities concerning gender roles in SET.

1 Gender related perceptions keep girls out of SET

Chances for girls growing up in Germany to engage in science, engineering and technology (SET) remain significantly lower than those for boys: Even today, less than 20 per cent of the engineering degrees are granted to women, and many of these did not grow up in Germany but have a migration history that exposed them to the educational system and cultural background of different home countries[46]

A common result is that educators at schools and universities as well as significant persons in the working environment are key to the development of self-images as well as education and career choices of adolescents. It is therefore important for any programme trying to improve the situation to provide insight and training for these significant persons in the gender specific aspects of their interaction with students and subordinates.

---

46   We refer to personal experiences from more than ten years teaching and being faculty members of engineering departments at German Universities of Applied Sciences.

1.1 Socialisation and role behaviour

Growing up in any society is intimately related to socialisation, i.e. a person's acceptance of norms and values of that society as "naturally given". The socialization messages that individuals receive in the social, cultural, organizational and family environments surrounding them form a set of stereotypes that strongly influence – and simplify – their decision making processes. This also includes preconceived notions on the expected behaviour of members of the society in particular situations. This expected behaviour is named role behaviour. The expectations concerning the specific behaviour of men vs. women are referred to as gender roles, which are accompanied by prevailing "gender stereotypes".

1.2 Gender stereotypes and career choices

As we will show in this paper, gender stereotypes play a particularly strong role in the development of training and career choices of girls. While boys interested in SET careers are typically considered within the norm and society's expectations, and are encouraged in their choice, girls find themselves confronted with messages of surprise and doubt and the need to justify and defend their preferences.

To create images of ourselves we constantly are dependent on others. Individuals and groups of persons on whom we depend more strongly than on others influence our beliefs, attitudes and mind sets in a most sustainable way. It is therefore of special interest to assess the degree educators at school and university as well as role models in the media, and other environmental aspects influence the self-image and choices made by girls and young women.

2 *Educators at school and university*

Researchers agree upon the observation that girls usually lose interest in SET around the age of eleven, i.e. at the onset of puberty, when physical gender differences become obvious and distraction due to stereotyped female role expectations is particularly high. In Germany, this is the time when SET education starts at school, giving girls a particular disadvantage right from the start.

It is during this time, that girls "learn" that high performance in SET subjects does not make them attractive to male classmates (Kessels & Hannover 2003).

2.1 Different teacher behaviour towards students of both genders

Whereas boys are encouraged in their interest in SET, girls with the same interests receive multiple signals that they do not meet the role expectations associated with them and consequently develop more defensive attitudes. For example, it has been observed that boys take and receive more room, time and teacher attention in robotic courses compared to girls. As boys were allowed to present their robots at the centre of the room and interfere with girls' presentations, the girls withdrew towards the fringes (Wiesner & Schelhowe 2004).

Results of questionnaire and interview studies show that a considerable number of female as well as male students observe different teacher behaviour towards students of both genders (Neuhäuser-Metternich & Krummacher 2007):

- At school, more girls than boys report that male teachers do not respect them, make more degrading remarks about them, and discourage and ridicule them more often than boys; more girls than boys do not feel acknowledged for efficiency by teachers of both genders.
- Having surmounted these hurdles and adverse experiences, young women who do carry on to a university education in SET are likely to experience further devaluation: They report feeling less credited for efficiency, discouraged due to degrading remarks and ridiculed by male university teachers; they perceive male university teachers giving more respect to male students – a behaviour that is also recognized by 16 per cent of male students; and they perceive male university teachers as giving more credit for efficiency to male students – a behaviour that is also recognized by 33 per cent of male students.

Adverse effects of gender stereotypes on teacher behaviour have also been revealed:

- In German elementary schools, where girls are getting lower grades in math and science subjects compared to boys, whereas boys are getting lower grades in German language classes; male teachers are generally giving lower grades to girls, but both boys and girls evaluated contact with male teachers towards students less positive than that of female teachers (Pfeiffer & Baier 2009).
- In German high schools, where girls are getting lower grades for the same performance as their male classmates and better grades in physics if the gender of the student is undisclosed to the teachers judging the exam (see Wiesner 2004).

2.2 "Nice girls don't ask"

A particular form of student teacher interaction can be observed when students are asking questions. In Germany, students of both genders complain that questions are not welcome or even not allowed since teachers do not want their courses to be interrupted. For female students the situation is often aggravated by the stereotype that "nice girls don't ask" (Babcock & Laschever 2003, 2007). Therefore, when Berlin University of Applied Sciences and Technology and Economy introduced a new single-sex educative course in Computer Science and Economy For Women, the fact that "[s]tudents are allowed to ask questions" was a particular marketing feature[47].

Different interaction behaviours with teachers of boys and girls have been observed in special robotics-courses, when girls more often than boys addressed the teachers with questions and were thus drawing more attention and getting more influencing remarks. Teachers in return more often suggested to girls to follow the written directions, and thus implicitly discouraged them from exploring their own solutions. This in turn may be interpreted by girls as if teachers had less confidence in their problem solving abilities. Since boys happened to ask questions less often, they were left on their own and were working on a more individual level. This in turn provoked teachers to further encourage boys in figuring out their own solutions (Wiesner & Schelhowe 2004).

Systematic analysis of gender differences in the OECD's PISA tests results of 15-year-olds (OECD 2009) shows that students are following their own gender-related perceptions. Statistical patterns seem to mirror the motivation and attitudes of students when in 32 out of the 40 countries and economies assessed, females tended to report lower mathematics-related self-efficacy than males, while males tended to have a more positive view of their abilities than females; the latter experienced "significantly more feelings of anxiety, helplessness and stress in mathematics classes" than males (OECD 2009: 21).

So while boys and girls show identical ability overall in problem-solving at age 15, girls fall behind when it comes to solving problems in a mathematical context. The study attributes this difference to the ways in which problem-solving tasks are contextualised in school mathematics.

3   Self-fulfilling prophecy and stereotype threat

Reality can be influenced by the expectations of others, as experiments on self-fulfilling-prophecy show. This influence can be beneficial or detrimental depending on which label an individual is assigned. The effects of teacher

---

47   See: http://fiw.f3.fhtw-berlin.de [last checked April 2009]

attitudes, beliefs, and values, on their expectations from students have been tested repeatedly. The Pygmalion effect or Rosenthal effect (Rosenthal & Jacobson 1992) more commonly known as the "*teacher-expectancy effect*" refers to situations in which some students perform better than others simply because they are expected to do so. In the case of female students in SET, the teacher-expectancy effect leads them to internalise the expectations of not being interested and talented. Teachers may unconsciously and unintentionally send signals about what they expect and thus behave in ways that encourage some students, e.g. male students in SET, and discourage others, e.g. a considerable number of female students in SET.

Stereotype threat is a phenomenon in which the activation of a self-relevant stereotype leads people to show stereotype-consistent behaviour, thereby perpetuating the stereotypes (Steele 1997). Belief in the validity of the stereotype is not a necessary condition for the threat to actualise, as long as the stereotype is known by members of the marginalized group (Steele, Spencer, & Aronson 2002). Women's mathematical performances decrease when their gender is made salient (O'Brien & Crandall 2003; Brown & Josephs 1999). This underperformance effect was the same even when privacy of test outcomes was suggested, so that participants did not need to fear negative evaluation by peers (Inzlicht & Ben-Zeev 2003); but it was experimentally decreased when they were told to consider experiential accounts for underachievement in math (Dar-Nimrod & Heine 2006; see Neuhäuser-Metternich 2007a). Investigations like these show that stereotype threat can be reduced when people focus on the malleability of the traits at hand.

Neuropsychological studies gave insight into the brain mechanisms leading to the negative consequences of stereotype threat: With functional magnetic resonance imaging the neural mechanisms underlying women's underperformance in math were made visible. In situations, when women felt under threat, they showed the tendency to activate the so called "social" areas in their brains instead of concentrating on the solution of the math question (Krendl et al. 2008). Thus the fear being evaluated by other people is interfering with the ability to perform on optimal level, since girls or women are preoccupied with thoughts about reputation building like: "How can I make myself appear competent?"

3.1 Stereotype threat and self-image

Fears from stereotyping threat may interfere with self-image in SET, when girls at the age of 14 to 19 years see themselves as talented and interested, yet boys of this age estimate their talent and their interest as significantly higher than girls do (Neuhäuser-Metternich & Krummacher 2007).

As a further consequence of stereotype threat and the persistent experience of anxiety, female students in SET may protectively disidentify with academic achievement in domains they are perpetually told not to belong to (see Steele 1992). A German questionnaire study in which girls said not to like the word "technology" supports this hypothesis (Technikzentrum Minden-Lübbecke 2009).

Such academic disidentification has the negative effect of inhibiting scholastic success in the domain where it occurs. Low achievement in SET may then be counterbalanced by increasing identification with domains having better prospects, such as language or culture.

3.2 Official devaluation and stereotype of women's underachievement

In 1897, the year when Marie Curie had her first daughter shortly before starting her PhD work on her way to the Nobel prize, Max Planck replied to a survey on women at university:

"If a woman has a particular talent for theoretical physics, which is not often the case but happens from time to time, and if moreover she desires to develop her talent, ... I shall be pleased to permit her to attend my lectures and exercises on a trial basis, this permission being revocable at any time, inasmuch as this is compatible with academic order ..." and emphasizes " ... that natural laws can never be ignored without severe damage which, in the present case, would be seen especially in the next generation".

On 14 January 2005, Lawrence Summers, then president of Harvard University, still speculated that one reason why women are underrepresented in science and engineering professions is because of a "different availability of aptitude at the high end" (Lawrence H. Summers 2005). Reactions to this argument together with other complaints lead to his demission in February 2006; by then, however, he had strengthened a long standing stereotype.

In Germany, there are now at least some presidents of prestigious organisations like the Deutsche Forschungsgemeinschaft DFG (German Research Foundation), who openly advocate quota regulations "if we do not want to exclude 40 per cent of our intellectual potential" (Ernst-Ludwig Winnacker in 2006, then president of DFG), while others like the Fraunhofer-Gesellschaft and Max-Planck-Gesellschaft still see this approach as incompatible with the selection criterion of scientific excellence.

4 Mentoring to fight stereotype threat

By introducing mentoring schemes in schools, universities, and in the economy in Germany, new ways of designing educational policies that promote gender equity have been developed in the last twenty years. Mentoring is used to improve recruitment and reduce the number of students who drop out of studies as well as enhance the number of women in leading positions.

Moreover, with institutionalised mentoring programmes, women have reached more visibility, they are seen and heard as well in a more official way.

Thus by organizing mentoring programmes for women a broader discussion or conversation on the problem has been started – to change it from being a problem of the few to becoming a problem of the majority.

Mentoring has proved to be a very successful way to attract girls to scientific and technical studies and to support women in SET-careers. However, to build up the necessary perseverance and self-confidence needed to survive in a still male dominated science-world requires more than the attempt to change the ideas and attitudes of girls and women. It is the gender competence of the educators at school and university as well as other significant persons that needs to be improved to assure an assertive environment in which the talents of girls and young women can grow to full strength (see also Neuhäuser-Metternich 2010).

4.1 Mentors gaining gender competence

Mentoring is a support concept where an experienced person, the mentor, meets with a less experienced person, the mentee, to establish an open and trusting relationship.

The Ada Lovelace[48] Mentoring concept (www.ada-mentoring.de) e.g. is built on the belief that effective mentoring benefits both the mentee and the mentor – and the evidence supports this notion. The mentees receive a lot of information, and those who are already interested in SET are encouraged to choose a career in these fields.

But the benefits granted for the mentors seem to be at least as important (Neuhäuser-Metternich & Krummacher 2009):

- <u>Students as mentors</u>: The Ada Lovelace Mentoring-programme supports self-confidence, leadership skills, and interest in professional development of the student mentors. In addition, these students have the chance to make new friendships, build up their own network, and feel less isolated in their work environment as a result of the mentoring experience. They gain pride and satisfaction by mentoring, and the good relationship with other mentors has a positive impact on their professional performance. Most of them get sensitized to the special needs of female stu-

---

48 Ada Lovelace is mainly known for having written a description of Charles Babbage's early mechanical general-purpose computer, the analytical engine that never was built. She is today appreciated as the "first programmer" since she was writing programs for machines (that is, manipulating symbols according to a given set of rules). She also foresaw the capability of computers to go beyond mere calculating or number-crunching while others, including Babbage himself, focused only on these capabilities.

dents in engineering departments where they had denied these in the time they had been isolated.
- Educators and executives as mentors: When educators or executives act as mentors, these women and men learn a lot about the mentee, the young women and the reasons for their need of support. They experience that girls feel pressure to do everything and please everyone and are still discouraged from studying a scientific or technical discipline by various forces like:
- Pressure from peer groups to choose the subjects that the majority of girls are studying to avoid exclusion from female networks as well as losing attractiveness for male classmates.
- Science teachers approaching girls with teaching methods that are not appealing or stimulating, and not taking into account that the interests of girls are more diverse and not limited to the chosen major field of studies.
- Being objects of degrading remarks as well as experiencing widespread prejudice and gender stereotypical behaviour, e.g. amazement about their presence in the scientific or technical courses.
- Being confronted with opinions and myths about the inadequacy of women in the chosen scientific or technical field of studies.

5 Images and the media

Stereotypes about girls' and women's lacking of aptitude for SET are built up and strengthened at schools and universities on a more subtle level by interacting with teachers and peers as has been shown. But even if girls and women have been conserving their positive self-image, they need a huge amount of self-esteem in order to hold against the powerful male images in SET cultivated in the media.

5.1 Powerful male images – lack of positive female images

The EU funded project WOMENG used a cross-cultural approach to understand why so few women are still attracted towards technological education. From 2002 to October 2005 gender-specific cultural, structural, organisational, and individual barriers were investigated by seven partners across Europe. In Germany, 100 female and 100 male students from six technical universities were interviewed, the web pages of these organisations were analysed and the didactics in curricula of engineering courses were evaluated. Results indicate that the male image of science and technology at these universities is in strong contrast to the female image held by girls and young women (Sagebiel 2005).

"What happened to the image of science?", Londa Schiebinger (1989:148f.) asked. And in a historical review she stated:

"Though feminine icons were to remain in the broader culture (as representatives of liberty, justice, and so on), they disappeared from frontispieces of scientific texts.... During the 1800s explicit images of science are replaced by implicit and popular images of the scientist as an efficient male, working in a modern laboratory, most even wearing a white lab coat. Though this image is a literal one, it is also allegorical and points to a set of meanings that go beyond the literal image. The scientist is now an isolated individual, profoundly alone. Not evident in the picture are the props and crews that keep this man at the centre stage – his colleagues, his technicians and graduate students, his secretaries and even his wife. Absent too are the patrons and politicians influencing his work. This self-sufficient individual is of serious demeanour, and he is active. The fact that he is white and male is both descriptive and proscriptive; the image cultivates its own clientele."

This male image of science is a very powerful one in defining the person who is a scientist as well as what science is like; it gains even more power since there is no positive female image to counteract. In the media, at school, and at university the very fact is daily proved: "Scientists are male!" The OECD study therefore sees "a lack of role models for females (famous scientists, family members, etc.)" as one of the reasons for girls not choosing SET as careers in spite of given ability (OECD 2009). The problem is not easy to solve though, since the few women who have persevered in SET are quickly overloaded by too many requests to serve as role models and encourage others to follow them.

5.2   Positive female images in SET by video clips

Role models are presented by the media almost 24 hours a day. Most of these are representative of the stereotypes that prevail in society. It is only in recent years that a few projects have tried to counteract this influence, e.g. by associating science and technology with more feminine attributes. One example is the European organisation "Public Awareness of Science" (EuroPAWS), which awards prizes for the best transmitted television dramas bearing on science and technology and the most innovative factual and fictional drama. To "develop positive role models for young women rather than the usual clichéd stereotypes", they have launched the EU funded project "European Women in Science TV Drama On Message" (EuroWistdom) encouraging and supporting writers or producer-writer teams that show women in leading roles in science, engineering and technology based stories on TV[49].

The UPDATE project, also funded by the EU in the 6th framework programme, follows a slightly different approach based on the concept that choosing or rejecting a career path is linked to identity construction. A per-

---

49   See www.eurowistdom.eu, www.europaws.org, see ADA-Mentoring, 18, 2007: 29

son's interests and dislikes become part of a strategy to create and express one's identity. Girls and young women already active in SET should therefore be the best advisors in developing effective strategies to recruit and retain female scientists. UPDATE, which stands for "Understanding and Providing a Developmental Approach to Technology Education", therefore involves representatives of the target audience: Video clips and information material to attract girls towards science or engineering studies are produced by engineering students at Dortmund University of Applied Sciences and Arts in co-operation with the Ada Lovelace Mentoring Association and supported by the Federation of German Employer's Associations in the Metal and Electrical Engineering Industries (GESAMTMETALL).

Students worked together in a peer mentoring group, reflecting and documenting their own experiences made in the course of their studies. The videos contain both self-images of their producers as well as shared visions of the respective scientific communities. They aim at providing insight into this "strange species" of girls and women choosing a science career path – who are they, what motivated their choice, and how do they feel working in a SET-department and/or experiment-oriented institute (Neuhäuser-Metternich 2007b; 2009)?

5.2.1 Objectives of the video clips

Objectives emphasized in the video clips appropriate for the age-level (17 to 19 years old) of girls in the final classes of high school are:

- To develop role models for girls and young women.
- To de-construct gender stereotypes and promote technology as a relevant area for both girls and boys (Dakers et al. 2009).
- To support positive attitudes towards technology, and raise curiosity, awareness, interest and motivation.
- To promote knowledge and understanding about technological environment and solutions, materials, tools, equipment, concepts and classifications.
- To promote equal opportunities as well as gender balance by presenting female students acting professionally on the stage of technology departments at university.
- To allow recognition of technical problems to be solved.
- To allow recognition of problem solving skills of female students.
- To encourage creative problem solving, inventiveness and innovativeness.
- To provide experiences of success and accomplishment, thus strengthening self-esteem and empowerment.

- To provide opportunities to develop skills using tools and equipment, techniques, measurement, and planning.
- To provide self-assessment meaningfulness of the activity for learners, functionality of solutions, ethical perspective, e.g. environmental sustainability.

5.2.2 Three video clips

In the following three video clips not all of the above mentioned objectives could have been visualized in each of the clips. These clips were produced in order to offer the opportunity to reflect on classroom beviour as well as on other relevant aspects of the complex situation of girls and women in SET.

5.2.2.1 Izabella Vasileva, student of electrical engineering

This young student was fascinated by technology from early childhood. Originally coming from Uzbekistan, she compares her experiences made in school in an eastern country with those gathered currently at Dortmund University of Applied Sciences and Arts: She prefers the teaching atmosphere in Dortmund, in particular the freedom to ask questions about anything that remained unclear at any time.

5.2.2.2 Reducing noise by application of passive resonators – female students on search for a calmer environment

Tatjana Aust, Denise Exner, Swetlana Grebennikov, Julia Hilsmann, Rebecca Kalweit, Anne Kujat, Nathalie Latus, Johanna Uhl, students at the Department of Mechanical Engineering at Dortmund University of Applied Sciences and Arts, present a complex experiment in their acoustics laboratory, where noise and vibration in technical systems like passenger cars are reduced by application of passive resonators based inside doors or on carriage platform.

The young women explain their work in an interesting and stimulating manner. They let the audience participate in the research atmosphere of female students collaborating and communicating, and demonstrate that technology and science courses can provide fulfilling experiences to girls.

In addition to the video, the students produced an accompanying booklet giving readily understandable explanations for all relevant experimental and scientific details (Neuhäuser-Metternich 2009). Video and booklet together provide detailed insight into research life and are also useful as teaching material.

5.2.2.3   Sandra Stahlberg, diploma in engineering at the University of Applied Sciences, scientific assistant at the laboratory for Electrical Building Automation Systems and Machines at Dortmund University of Applied Sciences and Arts

Sandra Stahlberg, a young engineer talks about her career choice, her experiences first at university studying electrical engineering and then working with different employers, and finally her decision to return to university and work as a lecturer. She is presented together with her horse in her leisure time as well as in laboratory settings giving instructions to male students.

Video clips[50] are presented in schools as well as at different fairs, conventions, congresses, and other events and there have been reports on the project in several journals.

6   Conclusions and outlook

If gender sensitivity develops and participation of women in science and technology grows, the culture of science and engineering itself will change and the results will improve. Then there will be a greater chance that gender analysis will be done and will lead to gendered innovations referring to transformations in the personnel, cultures, and content of science and engineering. Gender analysis, when applied rigorously and creatively, has the potential to enhance science and engineering by sparking new perspectives, new questions, and new missions (Schiebinger 2008).

References

Babcock, Linda & Laschever, Sara. 2003. Women Don't Ask: Negotiation and the Gender Divide, Princeton University Press (see ADA-Mentoring 5, 2004, 24; ADA-Mentoring, 10, 2005, 29)
Babcock, Linda & Laschever, Sara. 2007. Asking for It: How Women Can Use the Power of Negotiation to Get What They Really Want, Bantam Books
Baumhöfer, Klaus. 2004. Die Lehre verändern, denn Nadine studiert nicht Elektrotechnik; in: ADA-Mentoring, 6: 22-23
Brown, R.B. & Josephs, Robert A. 1999. A burden of proof: Stereotype relevance and gender differences in math performance. Journal of Personality and Social Psychology, 76(2): 246-257
Dakers, John R.; Dow, Wendy & McNamee, Lynsey. 2009. De-constructing technology's masculinity. Discovering a missing pedagogy in technology education, in

---

[50]   The video clips are available on the following websites: http://www.fh-dortmund.de/de/studi/fb/3/personen/lehr/neuhaeuser-metternich/fue/koop.php
http://update.jyu.fi/index.php/Motivating_girls_with_films

International Journal of Technology and Design Education, Springer Netherlands, special edition dedicated to the UPDATE project, http://www.springerlink.com/content/102912/?Content+Status=Accepted

Dar-Nimrod, Ilan & Heine, Steven J. 2006. Exposure to Scientific Theories Affects Women's Math Performance. Science, 314, 435; http://www.sciencemag.org; ADA-Mentoring, 17, 2007: 12-13

Enders-Dragässer, Uta & Fuchs, Claudia. 1989. Interaktionen der Geschlechter. Sexismusstrukturen in der Schule. Beltz: Weinheim

Inzlicht, Michael & Ben-Zeev, Talia. 2003. Do high-achieving female students underperform in private? The implications of threatening environments on intellectual processing. Journal of Educational Psychology, 95, 4: 796-805

Kessels, Ursula & Hannover, Bettina 2003. Für Mädchen verboten? Klassenbeste in Physik ist bei den Jungen besonders unbeliebt. ADA-Mentoring, 4: 23-24

Krendl, Anne C.; Richeson, Jennifer A.; Kelley, William M. & Heatherton, Todd F. 2008. The Negative Consequences of Threat: A Functional Magnetic Resonance Imaging Investigation of the Neural Mechanisms Underlying Women's Underperformance in Math, Psychological Science, 19: 2: 168-175

Neuhäuser-Metternich, Sylvia. 2007a. Macht der Gene oder mächtige Institutionen – was verbirgt sich hinter unterschiedlichen Mathematik-Leistungen von Frauen und Männern? ADA-Mentoring, 17: 12-14

Neuhäuser-Metternich, Sylvia. 2007b. UPDATE für Technik-Bildung – EU-Projekt im 6. Rahmenprogramm, ADA-Mentoring, 17: 21

Neuhäuser-Metternich, Sylvia, Ed. 2009. Schallreduktion durch Resonatoren. Studentinnen forschen für eine ruhigere Umwelt. Begleitheft zum Film. ADA-Mentoring, 27

Neuhäuser-Metternich, Sylvia. 2010. Stop Overmentoring! Nur mit mächtigen Sponsoren werden Frauen die Chefetagen erreichen. ADA-Mentoring, 32: 12-18

Neuhäuser-Metternich, Sylvia & Krummacher, Sybille. 2007. Girls' and Boys' Perceptions of Science and of Science Teaching Practice UPDATE Progress Report Germany, WP 5, Partner 7, http://update.jyu.fi; www.ada-mentoring.de

Neuhäuser-Metternich, Sylvia & Krummacher, Sybille. 2009a. Ada Lovelace Mentoring – Engaging Girls and Women with Science and Technology, in: Tajmel, Tanja & Starl, Klaus (Eds.): Science Education Unlimited. Approaches to Equal Opportunities in Learning Science, Waxmann: Münster, New York, München, Berlin: 169-178

Neuhäuser-Metternich, Sylvia & Krummacher, Sybille. 2009b. Girls' and Boys' Perceptions of Science and of Science Teaching Practice; European Conference on Educational Research ECER 2009 "Theory and Evidence in European Educational Research" Vienna, Austria, 28-30 September 2009

O'Brien, Laurie T. & Crandall, Christian S. 2003. Stereotype threat and arousal: Effects on women's math performance. Personality and Social Psychology Bulletin, 29, 6: 782-789

OECD. 2009. Equally Prepared for Life? How 15-Year-Old Boys and Girls Perform in: School. Programme for International Student Assessment (PISA); http://www.oecd.org/dataoecd/59/50/42843625.pdf

Pfeiffer, Christian & Baier, Dirk. 2009. Lehrer im Urteil ihrer Schüler: Ergebnisse einer neuen repräsentativen Schülerbefragung, Kriminologisches Forschungs-

institut Niedersachsen. http://www.kfn.de/versions/kfn/assets/ lehrerbeurteilung-pfeiffer.pdf (13.05.2009)
Rosenthal, Robert & Jacobson, Lenore. 1992. Pygmalion in the classroom; Expanded edition. Irvington: New York
Sagebiel, Felizitas. 2005. Kulturen und Strukturen beeinflussen Ingenieurinnen in Studium und Beruf – Ergebnisse aus dem EU-Projekt WomEng. ADA-Mentoring, 12: 13-14
Schiebinger, Londa. 1989. The Mind Has No Sex? Women in the Origins of Modern Science, Harvard University Press Cambridge, Mass
Schiebinger, Londa. 2008. Gendered Innovations in Science and Engineering, Stanford University Press
Steele, Claude M. 1992. Race and the Schooling of African-American Americans. The Atlantic Monthly: 68-78, April
Steele, Claude M. 1997. A threat in the air: How stereotypes shape intellectual identity and Performance. American Psychologist, 52: 613-629
Steele, Claude M.; Spencer, S. J. & Aronson, J. 2002. Contending with group image: The psychology of stereotype and social identity threat. In: M. Zanna (Ed.): Advances in experimental social psychology, Vol. 34: 379-440, Academic Press: New York, NY
Summers, Lawrence H. 2005. Remarks at NBER conference on diversifying the science and engineering workforce, 14 January 2005. www.president.harvard.edu/speeches/2005/nber.html
Technikzentrum Minden-Lübbecke. 2009. Mädchen für Technikberufe gewinnen. Kriterien für erfolgreiche Aktionen und Angebote, ADA-Mentoring 28, submitted
Wiesner, Heike. 2004. Konsequenzenreiche naturwissenschaftliche Sozialisation von Jungen und Mädchen; ADA-Mentoring, 7: 9
Wiesner, Heike & Schelhowe, Heidi. 2004. Mit Robotik Chancengleichheit in der Schule fördern, ADA-Mentoring, 8: 27-28

Inspiring girls and boys in science, engineering and technology on virtual networks – the e-mentoring programme TANDEMkids at RWTH Aachen University

Marcel Lämmerhirt and Carmen Leicht-Scholten

Abstract

The mentoring programme TANDEMkids addresses male and female pupils of grades 6 to 9 (aged from 11 to 15) nationwide and offers an individual mentoring relationship. Mentors are students and postgraduates from SET subjects of RWTH Aachen University. The personal mentoring with university 'insiders' pursues the objective to provide access to SET subjects and related occupational fields and to sensitise for questions related to gender.

The key instrument and assistance tool represents an internet platform that is specifically constructed and adapted to the exchange of this mentoring relationship ("e-mentoring"). The following paper describes the strategic framework and processes to develop a coherent and sustainable concept for TANDEMkids. The focus is on the results of a web-based survey to identify interests and attitudes of the target group, as well as on the presentation of applications used on the platform. The main criteria are the long-term preservation of motivation.

1 Introduction

The mentoring programme TANDEMkids is part of RWTH Aachen University's Institutional Strategy which has been developed within the framework of the first funding of the German Excellence Initiative of the Federation and the German Research Foundation (2007-2012). The Institutional Strategy comprised 4 pillars: The pillar "Mobilising People" was dedicated to a sustainable human resources development. In this context offers and measures were developed for all target groups at the university, ranging from "school children to alumni". These activities set incentives for the occupation with SET subject areas and supported a scientific career at a technical university. Moreover another focus of this people policy of RWTH Aachen University was the establishment of gender and diversity related aspects on all levels of the university (cf. Leicht-Scholten 2009; Leicht-Scholten, Weheliye & Wolffram 2009).

The MINT Cooperation Programme within the pillar "mobilising people" had the overriding aim to inspire pupils for SET[51] subjects and to encourage them to study on subject in this area. Besides the mentoring programme TANDEMkids further measures were the mentoring programme TANDEMschool, the Pupil University as well as the Summer and Winter Schools[52]. Since autumn 2012 the MINT Programme and the constituent parts like TANDEMkids were consolidated as a permanent measurement of the university.

Through its incorporation in the Institutional Strategy of RWTH Aachen University, TANDEMkids is structurally integrated and should not be thought of as an isolated measure. At the beginning of the 6th grade the attention of pupils is brought to the various offers of RWTH Aachen University and they receive support during the study- and job decision-making process throughout their last years in school. Furthermore, the programme is organisationally connected to already existing mentoring programmes at RWTH Aachen University (cf. Leicht-Scholten 2007, 2008). The long experience and expertise of these other programmes in conceptual planning, organisational companionship and scientific evaluation generates positive impulses for the mentoring programme TANDEMkids.

TANDEMkids as a nationwide mentoring programme is aimed primarily at pupils from 6th to 9th grade (aged from 11 to 15), to provide them with a first glance into the diversity and the scientific range of SET topics and to support them when it comes to questions of school, study and career choice. Mayor importance is given to the individual and flexible supervision of the mentees as well as to a target group-oriented, gender-equal and interesting mediation. Mentors are university students and students on the way to a PhD in a SET subject at RWTH Aachen University. Specific seminars prepare the mentors for their tasks and clarify possible expectations of the pupils. In addition, pupils are prepared in a specific seminar in order to clarify the roles of mentee and mentor, as well as to optimise the communication process in relation to mentoring. The "tandems" – the couples of mentor and mentee – are generally matched according to the criteria of sex and technical-scientific interest. The exchange and the interaction of all participants take place through a specifically designed internet platform ("e-mentoring"[53]), which is meant to last for one year. A framework programme with visits and workshops at RWTH Aachen University also offers the possibility to experience the fields of activity at a technical university closely. The programme is ac-

---

51 The abbreviation SET stands for the subjects of Science, Engineering and Technology. It is used in Anglophone contexts in equivalent to the German term MINT (Mathematics, Informatics, Natural Science and Technology).
52 www.igad.rwth-aachen.de/engl/mint.htm
53 For detailed information about e-mentoring see Bierema & Merriam 2002 and Miller & Griffiths 2005.

companied by a scientific evaluation in order to identify efficiency, effectiveness and satisfaction.

Figure 1: The mentoring programme TANDEMkids

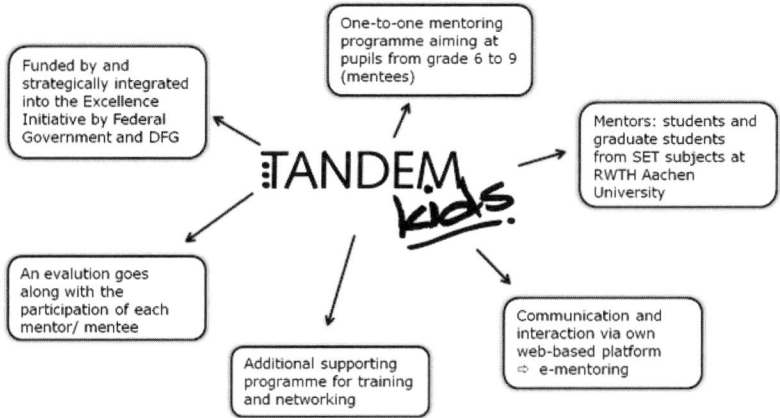

Source: Figure prepared by the author, 2011

In the following, the conceptual design and the practical implementation for the mentoring programme TANDEMkids will be presented. Three aspects are in the main focus:

- The pursued aims of the mentoring programme TANDEMkids and their theoretical and empirical foundation.
- The survey of a questionnaire during the conceptual phase among the target group of the pupils, as well as.
- The media-pedagogical ideas and their practical implementation on the internet platform as essential instrument of the mentoring programme.

The structure of the text is oriented on the mentioned aspects. Practical experiences and results of the evaluation cannot be delivered in the framework of this text, because the programme started in September 2009 and the registration of the pupils is still running. At the end of 2009 up to 50 and in November 2010 up to 80 mentoring relationships "Tandem-couples" had been established[54]. This constitutes a size that allows for an adequate coordination and supervision of the tandem-couples.

---

54 Approximately 45% of the pupils are female, the average age is 13.

2 The objectives of the mentoring programme TANDEMkids

The description and explanation of the theoretical assumptions as well as empirical results that form the basis of the conceptual planning of the mentoring programme TANDEMkids are orientated themselves by the defined objectives that the programme pursues. The objectives are consecutively evaluated by the dimensions of effectiveness, efficiency and degree of satisfaction concerning the realisation of the programme.

The following main objectives of TANDEMkids are presented:
- The increase of enthusiasm for SET fields,
- the sensitisation for gender aspects,
- the individual support of mentoring and
- benefits and social competences for mentors.

2.1 Inspiring high school students for SET

The main objective of the programme is to offer high school students an interesting and individual access to SET subject areas, which are appropriate for their age. The participating high school students can deepen their interests or discover new interesting subject areas for themselves. They receive advice to subject-related questions and also get consultation and support to receive the first understanding for the processes and characteristics of the study in generally and in specific courses of studies. Furthermore, the programme introduces them with the variety and creativity of possible fields of occupation in the SET areas to diminish negative clichés in the perception of certain professional categories and to close gaps in knowledge.

The background of this objective is, on the one hand, the acute shortage of young people in technical-scientific fields of work (1) and, on the other hand, the early formation processes of occupational interests (2).

(1) Today already ten thousands of skilled employees such as engineers are absent in Germany; that causes an immense economical damage and slows down the economic growth. Forecasts draw sombre scenarios here that have to be proactively faced especially in times of a worldwide economic crisis (Stifterverband für die Deutsche Wissenschaft 2009). For this reason RWTH Aachen University has set the goal to actively pursue a sustainable and efficient promotion of young people starting already during early years of school education. Thereby, RWTH Aachen University not only aims at quantitatively raising the number of its prospective students but also adequately preparing students for their studies in order to minimize wrong expectations and to decrease the drop-out rates (RWTH Aachen 2009).

(2) Studies have shown that children (until the age of 13) orientate their occupational images according to the social recognition of single occupations (cf. Taskinen/ Asseburg & Walter 2009: 80f). After that a reorientation phase begins focusing more on self-perceived interests, personality traits and wishes. Realistic occupational images are developed by the age of 15/16 years but are not firmly established in their orientations, however, they can indicate a trend (cf. ibid: 80f). Girls conversely start to have a more exact occupational image earlier than boys. Decisions of career choice are influenced by the self-assessed competence of the pupils for a certain occupation that is generally influenced by grades and their overall performance in class. However, high school students do not always have realistic assessments of the competence profile of a certain occupation. Accordingly, the interests and the intrinsic motivation for an occupation or a subject are equally important for the career choice than the development and strengthening of a competence self-assessment (cf. ibid: 81).

2.2 Sensitisation for gender aspects

Next to the increase of interest and enthusiasm for SET subjects, a third defined objective is the sensitisation of participants for gender equality. Girls and women face prejudices as well as implied and open discriminations in technical and scientific areas of work in particular. Normally these mechanisms are effective in socialisation and interaction processes like the classroom (cf. Faulstich-Wieland, Weber & Willems 2004). Through individual support in the mentoring programme these processes of social exclusion can be counteracted.

Girls show negative attitudes and perceptions regarding SET subjects that are based on unfavorable and negative experiences and manifest themselves in emotional distance and restraint towards technology and natural sciences in general (cf. Hannover/ Bettge 1993: 15ff; Faulstich-Wieland, Weber & Willems 2004: 9ff; Hannover/ Kessels 2004: 52ff).

According to the model of "latent components of perceptions for the differentiation of forms and effects of patterns of attitudes" (cf. Zanna and Rempel 1988) it is essential to promote girls in the development of so-called evaluative and affective attitudes.

Evaluative attitudes can be defined with the intention model by Fishbein and Ajzen (1975): An action is preceded by an intention that is predictable from the manner (sum of all consequence acceptances of an action and their assessments) and the subjective norm (sum of the images and expectations towards the most important persons and their assessment/ weighting). Manner, subjective norm and intention form evaluative attitudes. Girls have often negative perceptions about and subjective norms towards boys that can result in a lower interest for SET occupations; thus, girls comparably expect more negative consequences and less social recognition if they decide to pursue a career in a SET occupation. Hannover and Bettge (1993) further developed the intention model of Fishbein and Ajzen and define three areas that determine affective attitudes: Model influence (1), future images (2) and achievement motivation (3). There are no female role models who attain social recognition (ad 1). Increased expectations in qualification profiles on the job market do not take into account the gendered structure of certain professions. Most girls and women choose with an eye to their own future professional prospects and homes lives that allow a reconciliation of work and family life (e.g., a teacher). Studies clearly prove that the importance of family ties and partnerships is higher than the realisation of professional aims (cf. HISBUS 2008) (ad 2). Achievement motivation can be described as a motivation depending on solution probability of a task and the incentive for the successful solution. Thus, the anticipation of success or failure describes the achievement behaviour as a result of an emotional conflict: If SET areas are considered as "male-dominated areas", girls anticipate a failure in the solution. In this context positive results for girls are considered more as luck, exception or even as cheating (ad 3) (cf. Hannover/ Bettge 1993: 20ff).

Girls receive a role model by mentoring that shows them how to oppose obstacles, barriers and discrimination in the development of own interests. Furthermore, the relationship can optimise decision-making processes for future education phases; e.g. consciousness can be raised that a female high school student decides to attend an obligatory elective subject in a SET field, although only few girls are participating yet. Mentees get a role model that is normally missing in their social environment due to the gender-specific division of work. In mentoring relationships evaluative and affective attitudes are being addressed through the encouragement of young girls to engage in SET subjects. In doing so a positive self-image is developed and trust in own achievements and capabilities is built.

It is crucial not to force decisions and opinions upon the girls, but rather to give them impulses and opportunities for their future careers. Eventually girls should decide for a study subject that relates to their talents and not societal expectations.

2.3 Individual support by mentoring relationships

The methodical application of the mentoring relationships creates an adaptable and individual possibility of support for the participants (cf. e.g. Franzke & Götzmann 2006; DuBois & Karcher 2005; Jekielek et al. 2002). Thus, not only the personal interests and concerns can be addressed, but also other specific competence areas can be extended. It is essential that this process is performed in an appropriate manner for the target group. Otherwise, it could cause excessive demands on the mentees and demotivate them for further participation.

Specific competence areas, in this context, mean the perception and acknowledgement of own weaknesses and strengths, the formulation of realistic objectives and possibilities for obtainment as well as an increase of self-confidence.

Mentors deliver an additional assistance since they build up a trusting and amicable relationship with the mentees and can give them important impulses. Mentors answer specific questions and help in learning processes, however, give no orders or moral instructions. They rather support the mentees to break new grounds, give constructive feedback and try to keep an eye on the fulfillment of the objectives of the mentoring relationship.

The involved mentors are prepared with the help of a special training course. Additionally, they receive work materials and further assistance on the internet platform of TANDEMkids, which they can also use for the exchange in the mentoring relationship.

2.4   Aims for mentors and female mentors

The participating high schools students are not the only ones to profit from the programme, the mentors also gain a lot (cf. e.g. Franzke & Götzmann 2006; Richert 2006; DuBois & Karcher 2005). Mentors can discover new perspectives as well as new insights for their own career prospects. Mentors run retrospectively through their own educational pathway and can reconfirm their career decisions retrospectively. Furthermore, their communicative and leadership competences are further developed through the communication with a younger target group. In the conveyance of scientific-related contents the criteria of comprehensibility and the focus of the target group is acknowledged.

Teaching students/Students of pedagogy can get credit points for the participation in the programme. For doctorate students the possibility exists to certify the participation as a competence acquisition of key qualifications for the doctorate supplement[55] in accordance with the respective responsible person. An expansion of the certification of the participation with TANDEMkids for relevant study fields is planned.

An individual questionnaire was carried out alongside an analysis on existing empirical results. The questionnaire supported the step of the conceptual elaboration of the objectives to the practical conversion, but ensures concurrently the consideration of needs and inquiry trends of the target group.

---

55 The doctorate supplement at RWTH Aachen University certifies the acquisition of specific key skills to every doctoral candidate beside the professional qualification.

3 Questionnaire evaluation

In the concept phase of the mentoring programme TANDEMkids, a questioning of the target group was carried out in order to observe a trend in attitudes, experiences and preferences concerning technical-scientific subject areas and knowledge on the application of the computer. The reason for the decision to develop a specific questionnaire lies in the fact that is essential to draw a sufficient picture of the target group. For example, our questions are aimed at adequately covering the topics of career choice, study and interests in SET faculties. Furthermore the age group should be specifically addressed to survey the applicability of computer and internet. The advantage of an individual questionnaire also originated in the idea to ascertain expectations and prospect on the topic mentoring and to obtain information on predilections regarding graphic presentations. The higher aim of the questioning consisted in the development of a basis for the conceptual design of the programme that is oriented towards the needs of the target group and speaks the "language" of teenagers, e.g. through sketching information such as flyers and the design of a target group-oriented web page.

The questioning was conducted as an anonymous online questionnaire which corresponded to the planned web-based realisation of the mentoring communication. As an instrument the open source product "limesurvey" was used (www.limesurvey.com). The programme has proved to be helpful and is now used in the web-based evaluation of the programme as well.

The questionnaire includes the following dimensions and suitable items:

- Personal data (gender, age, school form, class step).
- Extracurricular and interests at school (favourite and disliked subjects, other activities at school, leisure activities).
- Purposes and prospects of study (intention to begin a study, choice of the subject and reasons, bias towards academic studies).
- Choice of career (career aspiration, sources of information to occupational fields, self-assessment of strengths and weaknesses, choice of career criteria).
- SET (contact with technical devices, assessment of occupational fields in SET).
- Mentoring (willingness to communicate with students, possible contents of an exchange).
- Computer and internet use (Temporal and purpose-engaged behaviour of utilisation, access possibilities, communication behaviour on the internet, criteria for the assessment of internet sites).
- Graphic preferences (assessment of given patterns for internet platform, advertisement poster and logo).

Before the real start of the survey a pretest was conducted. The pretest ensured that questions and interrogative constructions were created comprehensively and fit into the expected time slot. Methodically it was carried out by means of standardised pretests with 7 test persons as well as a cognitive proceeding of the Thinking-Aloud-Method with one male and one female test person (cf. Lewis & Rieman 1994). The test persons were male and female pupils belonging to the target group of TANDEMkids and delivered several results for the improvement of the questionnaire.

The online questionnaire evaluated by the pretest was publicly announced by means of press release and information letters addressing schools which are in contact with RWTH Aachen University. The survey took place in June 2009 during a time span of 4 weeks. All in all 144 records with answers from pupils could be collected. Although this does not prove to be a representative picture of the target group, nevertheless, it delivers a first trend and therefore a basis for further development of the conceptual design of the programme.

In the following sections the most remarkable and most expressive results for the further programme design are outlined.

3.1   The sampling

All together, 144 records could be generated within the scope of the online questioning. 121 of them were filled out more or less completely, while 23 records were filled incompletely and therefore not integrated into the evaluation.

In total 70 boys and 51 girls took part in the online survey. The average age of the interviewees was 13.47 years. Most of the interviewees attended a *Gymnasium* (approx. 60%), but pupils of *Gesamtschulen* also took part (approx. 16%), also *Realschulen* pupils (approx. 4%) and pupils from *Hauptschulen* (approx. 12%) took part (8% were not stated). Most of the pupils came from North Rhine-Westphalia. However, due to contacts of RWTH Aachen University to schools all over Germany few interviewees came from Schleswig-Holstein, Baden-Württemberg, Lower Saxony and Rhineland-Palatinate.

The following aspects of the results will be presented:

- Attitudes to study and university,
- attitudes to mentoring and
- computer use and self-perception of technological competence.

3.2 Attitudes to study and university

Concerning the question of future plans after high school graduation, it became obvious that many of the pupils had the concrete plan to begin a study. Just 40% of the interviewees wanted to study at a German or foreign university, approx. 16% would like to finish a vocational education, approx. 13% preferred to take part in a binary system of study, and 9% wanted to take up occupational activities. Approx. 18% were as yet undecided. In respect to the type of school the results confirm that the permeability because of the German education system is limited. While pupils of *Gymnasien*[56] plan to enter university, interviewees of *Gesamt-*, *Real-* and *Hauptschulen* prefer vocational education studies or at least to study in a binary system. Besides, some interviewees are still indecisive about their career plans for the future. Concerning the question of a certain subject of study in particular, it was remarkable that only half of the pupils interested in studying could state a subject that they intend to study. In addition, art and social sciences pupils also named SET subjects as possible subjects to study particularly mathematics, chemistry, physics, biology and informatics.

With regard to the TANDEMkids programme, the results of the online survey confirm the need that pupils from *Gymnasien* should be better informed about the possibilities and range of studies at universities, in particular regarding SET subjects that specifically apply to technical faculties that are not represented as subjects in school Regarding *Gesamt-*, *Real-* and *Hauptschulen* possibilities and the opportunities to study these in general have to be clearly communicated to the pupils. This is particularly important because pupils still have the possibility to change to other forms of school in the 7th and 8th grade.

3.3 Attitudes to mentoring

Among the interviewees a distinctive readiness concerning the participation towards a mentoring-relationship (approx. 80%, N = 74) could be recognised. Typically, interviews involved exchanges where interviewees talked about experiences about a concrete study programme from their mentors. Typical questions referred to the everyday life of studying, the conditions for entry and strains of specific studies, or of study possibilities, and which difficulties can come up. Furthermore, interests of the pupils focused on criteria for decision-making processes, possible occupational fields of specific studies, as well as support with school duties, and future plans of the mentors.

---

56 For a general overview of the German education system take a look at www.bildungsserver.de.

Those that rejected participation in a mentoring programme mainly indicated having no interest in study or specific support as their reasons.

On the one hand these results fall into the preparation of the mentors concerning anticipated expectations as well as interests of the mentees referring to the mentoring relationship. On the other hand the public presentation of the programme emphasises the use and chances of the mentoring concept to reach a wider range of the target group in the future.

3.4   Computer use and self-perception of technological competence

To plan the development and configuration of the internet platform as an important tool of mentoring ("e-mentoring"), it was surveyed how the interviewees use computer and internet. More than 50% of the interviewees use the computer on a daily basis (N=104). Approximately 10 e-mails or message items are written in personal communication systems (e.g. in SchülerVZ.de) on an average per week (N=87). Next to subjective criteria for good web pages (e.g. handling, own creation possibilities or intelligibility) also good practice examples of successful web pages were collected.

In conclusion it can be stated that the interviewed youngsters use the computer for different purposes also, such as games, searching for school assignments, or communication at least once a week. These specifications of the evaluation formed a basis for the creation of the web-based platform of TANDEMkids, which is introduced in the next section.

4   *Design and Application of the Internet platform*

An essential instrument of the mentoring programme TANDEMkids is an individual web-based platform that provides a protected space for interaction between the participants. Furthermore, the platform facilitates organisational planning and makes extensive materials and activities available for employment and collaboration.

As a platform framework the open source product "moodle"[57] is used. "Moodle" is a learning management system (LMS) that has been in use since 1999 and is continuously developed (cf. www.moodle.org). It has been confirmed as reliable, flexible and capable of adaption in many contexts worldwide. The single functions from "moodle" can be used context-dependently and create a leeway that makes the application of an LMS for a mentoring

---

57 A short but good introduction to "moodle" provides an amusing presentation on the web pages from "slideshare": http://www.slideshare.net/scheppler/moodle-erklaert-mit-lego-presentation [05.01.11]

programme possible. Furthermore, the open source principle allows the extension and change of functions as well as adaptations in the graphic surface.

A "moodle" system provides the following functions (cf. Cole & Foster 2007; Höbarth 2007):

- Thematic or temporal arrangement by so-called courses
- Extensive privilege management for participants
- Supply of materials and information
- Synchronous and asynchronous communication tools like chat, forum, weblog and personal messaging system
- Organisational tools like calendar or votes
- Activities for cooperation and collaboration like Wiki pages, creation of podcasts, mind maps or virtual classrooms
- Comprehensive reporting and utilisation statistics

The following examples show how the internet platform can be used for the realisation of the mentoring programme TANDEMkids. The platform represents an tools to help lay the foundation-stone for the successful progress of a mentoring relationship. The web-based platform programme needs coherent and exact planning and monitoring to push targeted and motivated learning processes. This is supported by the accompanying scientific evaluation as well as sporadic polls that investigate demand-oriented needs, interests and wishes to be able to clarify the motivations and incentives for the materials and activities offered on the platform. Materials, information and TANDEMkids news are edited and prepared in an online-magazine that allows the participation of the mentees who can include articles or reports of their own.

The offered activities and duties are processed in a didactical way so that they are adapted to the target group concerning complexity and extent. They represent a voluntary activity that should be included in the mentoring relationship in order ensure inclusion of additional motivation. In no case is there any focus on competitive assessments of the results of the applications. In fact, the focus is on a comprehensive engagement with SET subjects, occupational fields and gender roles and the supply of interaction possibilities.

Wiki pages and various discussion forums provide different activities to explain the programme, these tools are briefly explained before an introduction of application examples is presented.

4.1   Wiki pages

Wiki pages involve a hypertext system of web pages whose contents can be extended, changed and discussed by the users. This allows an open learning setting that supports collaborative working forms.

The material structure of the contents is illustrated with the help of the Wiki pages. The perpetual uncovering of relations between the material structures relates the contents to each other. The relations are created by hyperlinks and in this way produce a steadily growing semantic network. An example of the use of Wiki systems is the online encyclopaedia Wikipedia (cf. Abfalterer 2007; Moskaliuk 2008).

4.2   Discussions forums

A discussion forum characterises an asynchronous communication tool that is not chronologically bound. In a forum questions or news are posted that other users can answer or can place remarks on. Thus, parallel processes of discussion and formation of opinions are generated, so that the users can reflect on contents and can process the single forum contributions independently on a deeper level (cf. Abfalterer 2007).

4.3   The application possibilities

The following examples of how Wiki pages and forums can be applied are put in the context of the objectives described in section 2. By applying certain practices the following aims should be achieved: Focusing and enlarging the interest in SET subject areas, reducing gaps in knowledge and prejudices concerning technical-scientific courses of studies and occupational fields, and raising awareness for gender equality in school and everyday life. Besides, the application of activities is picked out as a central theme as clearly as possible in the respective mentoring relationship; e.g. mentors draw the attention to activities, provide assistance, give thought-provoking impulses, and take part in the discussion of the results.

As examples of the applications achieved using the internet platform in TANDEMkids, the following concrete activities can be explained:

- Brainstorming: Collection of ideas
- WebQuest: Investigation tasks on the internet
- Text production
- Learning Blog/e-Portfolio: Every participant describes own experiences, open questions and unsolved problems

Brainstorming indicates the collection of spontaneous ideas that are structured and discussed afterwards. Brainstorming aims at activating, gathering and structuring the prior knowledge of the participants. As an instrument Wiki pages are particularly suited to brainstorming. Primarily at the beginning of the mentoring relationship, brainstorming is suited to identifying, for example, beliefs, images and prior knowledge of the high school students regarding certain occupational fields like engineer, certain courses of studies, or gender-specific role pictures in the SET subjects[58]. With the help of the results the mentors can set goals concerning the mediation of information and experiences in the mentoring relationship.

WebQuests are didactically prepared and structured web-based researching tasks that are used for single or team work phases (March 2004). The produced results in the form of articles are linked with each other and posted for discussion. As an instrument Wiki pages are also particularly suited for WebQuests. For example, research tasks can ask for the photographic representation of certain occupations on the internet. Empirical studies can prove that e.g. photos representing technical occupations on company web pages promote gender-specific discrimination of women in technical occupations and reinforce gender-related clichés in the choice of career (cf. Wächter 2009).

It is planned to use the application of text production for wording and publishing experiences in the mentoring programme, as for example institute visits or results of activities and tasks (e.g. WebQuests). The publication will be released in the form of a newspaper. With the help of Wiki pages it is possible to constitute co-operative working processes, the application of graphic creation options as well as the active interlinking of single articles.

Finally, the use of so-called Learning Blogs or e-Portfolios offers the possibility to be able to appreciate the level of experience, the perception of individual strengths and weaknesses as well as the gender-specific sensitisation of all participants (Lorenzo & Ittelson 2005). E-Portfolios are implemented with the help of a forum for every participant. Each e-Portfolio can be linked with other files and activities, e.g. with discussions in other forums or the results of a research task. Each e-Portfolio is protected against being viewed by other participants; however, it will be observable by the respective mentor. Thus every mentor can identify the level of development of the mentoring relationship, check the extent to which the objectives have been reached, reflect the present exchange, and give constructive feedback.

5 Conclusion and outlook

On the one hand, the mentoring programme TANDEMkids is integrated into RWTH Aachen University's Institutional Strategy and on the other hand organisationally into the mentoring concept of RWTH Aachen University. On the basis of this strategic framework, TANDEMkids addresses male and

---

58 For example, it can be asked how the pupils visualise a typical student or a student of a certain course of studies in the SET subjects. The mentioned gender-specific stereotypes and clichés can be used for discussion themes. Another example is the metaphor of mentoring as a cruise: mentees should decide what "luggage" they take along. This luggage represents the aspects of their mentor and the mentoring exchange they consider to be most important.

female pupils of the 6th to 9th grades (aged from 11 to 15) nationwide and offers them an individual mentoring partnership. The personal mentoring with students and postgraduates pursues the objective to provide access to SET subjects and related occupational fields and to sensitise for questions related to gender.

The key instrument and assistance tool represents an internet platform that is specifically constructed and adapted to the exchange of a mentoring relationship ("e-mentoring"). Didactically processed accompanying tasks and materials using forums, such as WebQuests via Wiki pages and e-Portfolios, increase the intensity and quality of the mentoring relationships. The stimulus for motivation and the fun factor represent the most important criteria while planning und using these applications. It is crucial to support and not to judge the dynamic and individual processes that characterise mentoring relationships.

The scientific evaluation of the mentoring programme TANDEMkids aimed for the measurement of the effectiveness, efficiency and grade of satisfaction in the mentoring relationships. It gave indicators and concrete criteria for enhancing existent applications and developing new ideas and processes. With its permanent consolidation the programme has become part of a sustainable human resource strategy of the university.

References

Abfalterer, Erwin. 2007. Foren, Wikis und Chats im Unterricht. Hülsbusch: Boizenburg
Bierema, Laura L. & Merriam, Sharan B. 2002. E-mentoring: Using computer mediated communication to enhance the mentoring process. Innovative Higher Education, 26(3): 211-227
Cole, Jason & Foster, Helen. 2007. Using Moodle: teaching with the popular open source course management system.2nd Edition. O'Reilly Media, Inc.: Sebastopol
DuBois, David L. & Karcher, Michael J. (Editors). 2005. Handbook of Youth Mentoring. SAGE: Tousands Oaks
Faulstich-Wieland, Hannelore; Weber, Martina & Willems, Katharina. 2004. Doing Gender im heutigen Schulalltag. Empirische Studien zur sozialen Konstruktion von Geschlecht in schulischen Interaktionen. Juventa Verlag: Weinheim, München
Fishbein, Martin, & Ajzen, Icek. 1975. Belief, Attitude, Intention, and Behavior: An Introduction to Theory and Research. Reading, MA: Addison-Wesley
Franzke, Astrid & Götzmann, Helga (Editors). 2006. Mentoring als Wettbewerbsfaktor für Hochschulen – Strukturelle Ansätze der Implementierung. Lit.: Hamburg
Hannover, Bettina & Bettge, Susanne. 1993. Mädchen und Technik. Hogrefe Verlag für Psychologie: Göttingen et al.

HISBUS. 2008. Kurzbericht Nr. 5. Kinder eingeplant? Lebensentwürfe Studierender und ihre Einstellungen zum Studium mit Kind. HISBUS Online-Panel. Hochschul-Informations-System GmbH: Hannover

Höbarth, Ulrike. 2007. Konstruktivistisches Lernen mit Moodle. Praktische Einsatzmöglichkeiten in Bildungsinstitutionen. Hülsbusch: Boizenburg

Jekielek, Susan M.; Moore, Kristin A.; Hair, Elizabeth C. & Scarupa, Harriet J. 2002. Mentoring: A promising strategy for youth development. Child Trends Research Brief, 21: 1-7

Leicht-Scholten, Carmen. 2007. Challenging the Leaky Pipeline. Mentoring in Engineering. Cooperation scheme between technical universities and research institutions for excellent female researchers. First results, in Gender in Engineering – Problems and Possibilities, edited by Ingelore Welpe and Barbara Reschke. Peter Lang Verlag: Frankfurt a. M.: 175-191

Leicht-Scholten, Carmen. 2008. Where is the Key to Success? A Comparative Evaluation of Mentoringprogrammes for Outstanding Female Scientists in Natural Science, Engineering, Social Science and Medicine, in Gender Equality Programmes in Higher Education. International Perspectives, edited by Sabine Grenz/ Beate Kortendiek/ Marianne Kriszio and Andrea Löther. VS Verlag für Sozialwissenschaften: Wiesbaden: 163-178

Leicht-Scholten, Carmen. 2009. Gender und Diversity im Mainstream der Wissenschaften – Wandel der Wissenschaftskultur durch die Institutionalisierung von Gender and Diversity Management an der RWTH Aachen, in Gender als Indikator für gute Lehre. Erkenntnisse, Konzepte und Ideen für die Hochschule, edited by Nicole Auferkorte-Michaelis/Ingeborg Stahr/ Anette Schönborn and Ingrid Fitzek. Budrich UniPress: Opladen: 41-52

Leicht-Scholten, Carmen; Weheliye, Asli-Juliya & Wolffram, Andrea. 2009. Institutionalisation of Gender and Diversity Managament in Engineering Education. European Journal of Engineering Education, 34 (5): 447-454

Lewis, Clayton & Rieman, John. 1994. Task-Centered User Interface Design: A Practical Introduction. URL http://www.hcibib.org/tcuid/ [05.01.2011]

Lorenzo, George & Ittelson John. 2005. An overview of E-Portfolios. ELI Paper, July 2005. URL: http://www.sorteoudla.org.mx/promueve/ciedd/CR/tecnologia/AnOverviewofEPortfolios.pdf [05.01.2011]

March, Tom. 2004. The Learning Power of WebQuests. Educational Leadership. December 2003/ January 2004: 42-47

Miller, Hugh & Griffiths, Marc. 2005. E-Mentoring, in Handbook of Youth Mentoring, edited by David L. DuBois and Michael J. Karcher. SAGE: Tousands Oaks: 300-313

Moskaliuk, Johannes (Editor). 2008. Konstruktion und Kommunikation von Wissen mit Wikis. Theorie und Praxis. Hülsbusch: Boizenburg

RWTH Aachen. 2009. Report 2009. Published on behalf of the rector by the RWTH Aachen University press office

Stifterverband für die Deutsche Wissenschaft. 2009. Nachhaltige Hochschulstrategien für mehr MINT-Absolventen. Edition Stifterverband 2009. Edited by Stifterverband für die Deutsche Wissenschaft and Heinz-Nixdorf-Stiftung

Taskinen, Päivi; Asseburg, Regine & Walter, Oliver. 2009. Wer möchte später einen naturwissenschaftsbezogenen oder technischen Beruf ergreifen? Kompetenzen, Selbstkonzept und Motivationen als Prädikatoren der Berufserwartungen in PISA

2006, in Vertiefende Analysen zu PISA 2006, edited by Manfred Prenzel and Jürgen Baumert. VS Verlag für Sozialwissenschaften: Wiesbaden: 79-106

Wächter, Christine. 2009. Challenging Cultures of Engineering – How words, concepts, and images (de)construct engineering as a male domain, in Women in science and technology. Creating sustainable careers, edited by European Commission. Office for Official Publications of the European Commission: Luxembourg: 69-81

Zanna, Mark P. & Rempel, John K. 1988. Attitudes: A New Look at an Old Concept, in The Social Psychology of Knowledge, edited by. Daniel Bar-Tal and Arie Kruglanski. Cambridge University Press: New York: 315-334

Mentoring programme TANDEMschool – encouraging young people to study SET subjects by a coherent MINT cooperation concept at RWTH Aachen University

Gehrt Hartjen and Carmen Leicht-Scholten

Abstract

The following shows how a mentoring programme, embedded in both the curricular structure of the students and the general Institutional Strategy of the university, the MINT cooperation concept in particular, can serve as a good practice model. It can also motivate pupils to study a scientific or technical (SET[59]) subject (or at least get more attracted to the area) as provide the participating students with important soft skills. The mentoring programme TANDEMschool addresses pupils of grade 11 to 13 (age 16-19) and offers them an individually chosen personal mentor (students and graduate students from the SET area at RWTH Aachen University). The MINT programme comprises of two mentoring programs (TANDEMschool and TANDEMkids), summer and winter schools for pupils, a pupils' university and doctorate scholarships on didactics as well as gender and diversity aspects in SET.

Introduction – starting point

Both the industrial and the academic sector agree that Germany faces a severe lack of qualified engineers and scientists. Several statements from leading national engineering networks (e.g. VDI, acatech) state that between 60 000 and 80 000 open positions exist for qualified engineers and the situation is similar in natural sciences and mathematics (acatech & VDI 2009). Due to the demographic development in Germany this situation is even going to worsen in the future and become a severe problem for the total industrial and economic situation, with even stronger effects than in other countries.

In order to solve this growing problem, the universities need to generate more graduates in these relevant subjects during the coming years and decades. For this purpose, more qualified and motivated young people have to be recruited and encouraged to study a SET subject. Another aspect is the significantly high drop-out rate among students of these subjects, which are, as

---

59 SET= Science, Engineering, Technics; also denoted as STEM (Science, Technics, Engineering, Mathematics)

surveys have shown, often caused by a lack of realistic information and a wrong image about the subjects, especially the structures, demands and typical work fields for alumni. It is obvious that it is not sufficient to simply increase the number of first-year students but rather their "quality", i.e. their information, motivation and preparation for the subject they are going to study in order to sustainably reduce the number of drop outs.

Regarding the universities, this problem cannot be approached by addressing only the students already present. They also have to focus on their future students to reach the predefined goals of the educational institutional strategy of RWTH (e.g. the university has set the commitment to guide 75 % of the students to a successful graduation before 2020 (RWTH 2009a). Thus, to recruit more motivated and qualified students for SET, it is necessary to address younger students and to foster their interest and fascination in subjects like mathematics and physics as well as technology. Hence, a positive image of these areas leads to a stronger tendency to choose a scientific or technical profession after school. This motivation has to be combined with realistic information, personal experiences and the presentation of role models, to avoid the development of a wrong image of the daily work environment in SET professions (e.g. about the duties of an engineer) or the study of a SET subject at university.

A further cause for presently unused potential is the severe gender imbalance in the SET area: Although 52 % of the students and graduates at German university are women, this applies to only 34 % of the students in SET subjects and only 22 % in engineering, without major changes in recent years (Breuer & Leicht-Scholten 2011; Statistisches Bundesamt 2008; Statistisches Bundesamt 2009). In technical universities this situation is even worse, taking a close look: At RWTH Aachen where 74 % of the students are in SET only 35 % of the students overall are women, with women make up only 22% within SET subjects and 19% especially in the field of engineering (calculations based on (RWTH 2009b). Hence it is evident that all measures to increase the number of qualified and motivated beginners in these subjects have to include a gender aspect and show a special effort in significantly increasing the number of female students, e.g. by presenting positive role models and special encouragement to choose a career in this male dominated area.

Mentoring as a strategic institutional and personnel development instrument

The observations and problems mentioned above are important for human resources development at any university or company, especially against the background of foreseeable demographic changes and shortages of skilled personnel in the future. In order to stay competitive universities have to attract and retain qualified employees and students. Evaluations have shown that many of the problems and needs mentioned above can be effectively addressed in a framework of mentoring.

Mentoring is an institutionalised and process-oriented strategy for human resource development in order to boost the career of junior scientists. In most cases, not only the mentees, but also the mentors will benefit from this relationship because the mentors enhance their advisory skills and their own personal networks as well as developing insight into the situation and self image of the junior scientists. At RWTH Aachen University, mentoring programmes are embedded in the institutional strategy of the university. The following mentoring concept is contained in its Mobilising People Policy as a part of the Institutional Strategy "RWTH 2020 – Meeting future challenges" (RWTH 2008) and is closely linked with other measures in the framework of this strategy.

The integral mentoring concept at RWTH Aachen University

RWTH Aachen University uses mentoring as an instrument of a gender equality approach to human resources development in order to specifically attract and promote high qualified female scientists on different career levels. One focus lies in the individual promotion of junior scientists in the natural and engineering sciences.

After establishing a programme for the support of female graduate and doctorate students in 2002, several further programmes for different target groups have been developed and implemented. The overall strategy of the university is based on the conviction that the diversity of its students and employees signifies a special opportunity for greater potential. Thus, the diversity of scientists on different stages of the career is considered as a positive contribution to the success of RWTH Aachen University. To ensure competitiveness and innovation of a university, the promotion of more female scientists is of special importance. Furthermore, in the light of this competitiveness, a university may not only concentrate on the support of junior scientists and students, but also has to make sure to attract and motivate qualified students and to provide them with an excellent education. The estimation of the latter at RWTH Aachen University can be seen in the recent Institutional Strategy for Excellent Teaching (RWTH Aachen 2009a).

At present, RWTH Aachen University has implemented a coherent concept for gender and diversity management that includes mentoring as a central element of gender-oriented human resources development in its Institutional Strategy. Hence, all mentoring programmes were coordinated and supervised by the Integration Team – Human Resources, Gender and Diversity Management (IGaD) (Leicht-Scholten 2008,2011). Situated at the rector's office, it supports and accompanies the university in this comprehensive process. The Integration Team has been the central stakeholder within the measure "Mobilising People", which is part of RWTH Aachen's Institutional Strategy (RWTH 2008) set in the framework of the German Excellence Initiative[60].

Figure 1: Mentoring as an instrument of career development

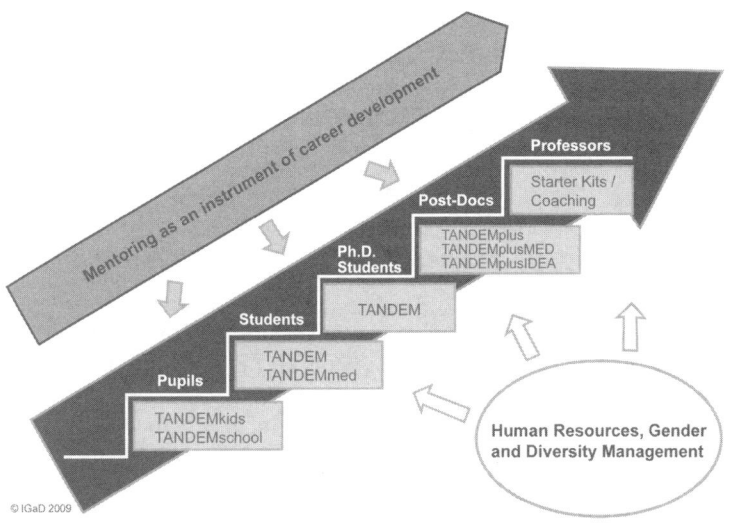

Source: IGaD 2009

RWTH Aachen University's mentoring concept currently comprises the following programmes (IGaD 2008):

TANDEMkids addresses pupils from grades 7 and 8 (age 12-15) to enhance their interest in SET subjects. The programme is mainly based on an internet platform (Lämmerhirt 2009).

TANDEMschool aimed at pupils from grades 11-13 (age 17-19) who are interested in studying a SET subject and offers them guidance to a secure and suitable orientation regarding their studies (see below).

---

60 For further information on the German Excellence Initiative, see http://www.dfg.de/en/research_funding/coordinated_programmes/excellence_initiative/index.html date when accessed

TANDEM addresses female graduate and doctoral students from all fields of study (except for medicine) and offers mentees support and guidance regarding future career decisions and qualification opportunities. TANDEM was the first mentoring programme at RWTH Aachen University established in 2002.

TANDEMmed started in summer 2008 and addresses female graduate students in medicine (from the 7th semester on) after their first preliminary medical, supporting and advising them in their individual career planning during their studies.

TANDEMplus is realised in cooperation between RWTH Aachen University and Technical University Karlsruhe. Since 2004 the programme has promoted female scientists on their way towards a professorship in natural sciences and engineering. The success of the programme has already been proven through a comprehensive scientific evaluation of the first three runs.

TANDEMplusMED addresses female postdocs, associate professors and habilitated scientists in medicine and natural science who are pursuing a scientific career at the medical faculty of RWTH Aachen University.

TANDEMplusIDEA included the development, implementation and scientific evaluation of a human resources programme for female postdocs of natural sciences and engineering. It is the first joint EU-project of the IDEA League (Imperial College London, TU Delft, ETH Zürich, RWTH Aachen and ParisTech), and the first international mentoring programme (Breuer & Leicht-Scholten 2009).

Starter Kits with additional **coaching** opportunities are a system of courses and seminars for newly appointed professors that offer training in areas like effective teaching, acquisition of funds and personnel management.

While the latter mentoring programmes focus on the career development of women in science, the two programmes for pupils with SET focus address both girls and boys; however, a proportion of at least 50% female mentees is intended in both programmes.

All mentoring programmes of RWTH Aachen University are based on **mentoring**, **training** (e.g. career building, leadership seminars) and **networking** modules. RWTH Aachen University offers the classical version of one-to-one-mentoring: In this version junior scientists are professionally and personally supported by one experienced mentor. Good personal contacts and functioning work alliances are in the forefront of a mentoring relationship. For the period of one year the mentors advise, give feedback, support the mentees in the development of potentials, competencies and strategies and motivate them in terms of their personal development.

The programmes also offer accompanying education modules (seminars, workshops, etc.) that are customised to the needs of the particular target group. All mentoring programmes comprise networking modules[61] in order to strengthen the network of the mentees in the scientific community. The mentors also have the opportunity to take part in the networks.

The joint objective of all mentoring programmes at RWTH Aachen University is to individually support the mentees and to encourage them to pursue a scientific career. At the same time the programmes take the specific requirements of the different target groups into account.

The MINT cooperation concept

The competitiveness of European universities will highly depend on the ability of innovation in science, quality standards in teaching and the establishment of the university in research and sciences on an international level. These goals can only be accomplished with the inclusion of diverse people who are involved in and promote new projects and ideas. RWTH Aachen has been meeting these complex challenges through the realisation of the pioneering "Mobilising People" policy that was formulated in the framework of the Institutional Strategy "RWTH 2020: Meeting Global Challenges" (RWTH Aachen 2008). The MINT Cooperation Programme has been an important part of this Institutional Strategy, and was funded within the framework of the German Excellence Initiative by the German Research Foundation. MINT is the German abbreviation for the fields of mathematics (Mathematik), computer sciences (Informatik), natural sciences (Naturwissenschaften) and engineering (Technik), and hence is used as an approximate equivalent of SET.

The main objective of the MINT cooperation concept has been to establish a new recruiting policy for the transitional phase from school to university to attract motivated pupils and encourage them to study natural sciences or engineering. Due to the special situation of SET subjects (see above), the concept incorporates a gender oriented approach with the aim of attracting more women into the respective fields, e.g. by presenting positive role models.

The integral concept consists of several interconnected measures that focus both on the transition from school to university as well as on the transfer of research results into school curricula. The following activities are part of the MINT cooperation concept:

- **Scholarships for didactics:** The programme offered 8 scholarships to doctoral students with a degree in teaching practice for didactical re-

---

61 Consisting of regular networking meetings for the mentees to support the connection they make amongst one another and in the scientific community, sometimes also with special invited speakers or external participants. Some of the programmes offer similar networking events and informal meetings for mentors.

search in the SET area. The fellowship programme is at the interface of recent research, further training of teaching staff and instruction. By means of these doctoral projects a joint venture of didactics in the different scientific faculties could be implemented with the aim to develop additional training modules for teachers. Further goals include promoting the connection between science and school and to attract pupils to natural science at the university in the long-term. The resulting modules, which offer didactic material for presenting selected items of recent applied research in school, were used for workshops in the Summer/Winter Schools and the Pupils' University. In the long term, an important effect will be to transport realistic and topical applications of recent research into the classroom and to strengthen the connection between university and schools.

- **Scholarships for gender and diversity aspects in natural science:** 10 scholarships have been offered to fund innovative, interdisciplinary dissertation projects in research projects that incorporate gender and diversity perspectives at RWTH Aachen University. The aim of these projects is, from the perspective of different subjects inside the SET area, to further a better understanding of the ways social factors can affect science.
- **Summer and Winter Schools for Pupils:** In cooperation with MINT-EC[62], an association of mathematical and scientific excellence centers at schools, RWTH Aachen organises a Summer and a Winter School per year for approximately 30 pupils who come from the network of MINT-EC schools (about 120 high schools with a focus on natural sciences in their school programme from all parts of Germany, including one German school in Istanbul, Turkey). In these four-day courses, the pupils from grades 11-13 take part in workshops on special areas of applied scientific and technical subjects not included in the regular school curricula and containing both theoretical and hands-on units. A special focus is placed on personal experiences with scientific matters, obtained in experiments and simulations in pairs or small groups, so that the participants can get an impression of both the study in the specific subject as well as of respective research activities. The programme for these summer and winter schools, during which all participants (including local participants) stay in a youth hostel, is enhanced by information about the study opportunities and demands of the subject within the workshop. Social events like sports, a guided city tour and dinners for the group complete the programme. The purpose of these additional activities is to promote networking among the pupils, the exchange of ideas and experiences and also the opportunity to discuss with scientists from RWTH Aachen University in an informal setting create sustained effects beyond the duration

---

62   Verein mathematisch-naturwissenschaftlicher Excellence-Center an Schulen e.V.

of the workshop. To prolong these long-term effects, the participants are offered the opportunity to take part in the mentoring programme TANDEMschool (see below). The content of the workshops for the summer and winter schools are mainly prepared by the holders of the didactical scholarships (see above).

- **Pupil[63] University:** Based on the experiences of the summer and winter schools, RWTH Aachen University offers in the framework of the pupil university during the summer holidays five-day courses in various SET subjects (in the first round in 2009 there was a choice of 11 different subjects). The aim of these workshops is to impart a realistic image of the structure, fascination, opportunities and demands of the chosen study subject to the participants, assisting them in choosing a suitable area of study and at the same time strengthening the connection between schools and university. Although the programme mainly addresses teenagers from the region, pupils from all parts of Germany and the rest of the world may also participate. In a combined approach, the workshop concepts developed for the summer and winter schools are also adapted to the pupil university.
- **TANDEMkids, TANDEMschool:** To implement long-term effects beyond the duration of a workshop, those activities are combined with the element of mentoring. While the programme TANDEMkids is explicitly described in Lämmerhirt 2009, the programme TANDEMschool is explained below. Both form a part of the MINT cooperation programme and cooperate closely with the other activities.

First results from the evaluation of Summer and Winter Schools

So far, three summer and winter schools and one pupil university have already taken place: In September 2008, a summer school on applied mathematics (workshops mainly about population dynamics and statistics) with 25 participants, in March 2009 a winter school on biology (workshops mainly about fermentation, green biotechnology and neurobionics) with 23 participants, and in September another Summer School on applied mathematics (workshops mainly on applications of orthogonal projection in sound and image processing, and statistic quality control) with 31 participants. In the summer holidays of 2009, the first pupil university took place with courses in 11 SET subjects (architecture, biology, chemistry, civil engineering, computer science, electrical engineering, material science, mathematics, mechanical engineering, physics, process engineering) with 260 participants.

---

63 The term «pupil university» refers to workshops and activities for high-school/secondary level students within university.

The following table shows a comparison of the gender distribution among the participants of these courses (the two schools on applied mathematics are combined), against the data for students and study beginners (for RWTH Aachen University and all German universities). The data for summer/winter schools and the pupil university are based on answers given in the evaluation.

Table 1: percentage of female participants/students (student data for winter sem. 2008/09)

	Summer/ Winter Schools	Pupil University	1st sem. students (RWTH)	1st sem. students (Germany)	Students (RWTH)	Students (Germany)
mathematics	46 %	53 %	49 %	52 %	42 %	48 %
biology	65 %	62 %	56 %	65 %	59 %	64 %
total/MINT	52 %	39 %[64]	23 %	37 %	22 %	34 %

Source: Data for RWTH Aachen University based on (RWTH 2009b), for Germany on (Statistisches Bundesamt 2008)

The data shows that the evaluated measures place female participation in the same range as the percentages of students or beginner students shown in these subjects This is not very significant, since the subjects mathematics and biology are (together with architecture) those SET subjects with the highest rate of female students, as the comparison with values for the whole SET area shows. Nevertheless, the remaining courses of the pupil university show that in those the percentage of women was significantly higher than among students of the same subjects. Since the male imbalance among students at RWTH is higher than in most German universities, it is necessary to put a special focus on attracting more women. Hence we see that such measures as summer/winter schools and pupils' universities are an effective measure to attract girls interested in SET subjects and foster their interest in this area. The evaluation of the summer/winter schools shows that most of the participants (75.6 % of the girls and 68.4 % of the boys) (probably) intend to study the subject of their workshop after school, while 62.5% of the girls and 81.6% of the boys answered that they would (probably) study a SET subject in general[65], which has partly been influenced by experiences of the workshop.

Although some pupils take part in more than one summer/winter school (on different subjects), the evaluation shows that most of the participants (90 %) had already chosen the subject in which the courses were embedded (mathematics or biology, respectively), as one of their "Leistungs-" or

---

64   in all 11 courses
65   Indication of strong or moderate interest on a five-valued scale

"Schwerpunktkurse" (chosen subjects with special focus and intensive courses in the German school system of grades 11-13 resp. 12-13).

TANDEMschool – a mentoring programme for pupils

Aim and target group

As we have seen above, universities and especially technical universities like RWTH Aachen University need to increase the number of graduates in SET subjects and hence the number of motivated and qualified students in these areas. Therefore, the mentoring programme TANDEMschool, as a part of the MINT cooperation concept, aims to attract pupils to the area of SET, make them discover their potentials and talents in this area and finally to motivate them to consider studying a SET subject after school.

The programme addresses pupils of grades 11-13 (age 16-19), with a basic interest in SET subjects, i.e. in the last two or three years of school (German "Gymnasium") before they obtain their "Abitur" and may start their studies. By participating, the pupils gain insight in the structures, contents and demands of certain study subjects and professions or career branches. Through active and personal support on several levels they are inspired and led to a well-based and suitable choice of study subjects and personal career options. To attain this goal and provide them with information and insights, their existing interest in the SET area is strengthened and increased by corresponding measures, information and practical offers.

Its special focus is to explicitly address young girls and to encourage them to consider a career in the field of SET, to increase the percentage of female students in SET subjects on a marked and sustained level and to improve equal opportunities. Programmes like these are especially capable of addressing young women, as the analysis of participants in measures like pupil workshops has shown (see above).

Although for most of the participants the interest in the SET area is coupled with outstanding success and good grades in scientific subjects (physics, chemistry, biology) and in particular mathematics, this is not a precondition for taking part in the programme. There are many reasons for low grades in a certain subject and grades often do not manifest a lack of talent or potential in the according subject, but rather a deficiency in motivation, little insight into the importance of the subject or simply the need for an alternative access to the field. Thus, these cases can be approached by personal inspiration and support of the programme, which might even lead to better marks.

The programme was at first only addressing participants of the MINT summer and winter schools (see above), it is now, however, open for pupils from all schools, mainly from the region around Aachen (including the adjacent Belgian and Dutch districts)

The mentoring

In the first place, the personal encouragement and motivation of the participants is reached by one-to-one mentoring. For a period of one year, each of the participating pupils is assigned a personal mentor (student or doctoral student from a SET subject or research area) to get an insight into the study of a preferred SET subject and to have a selected personal partner to exchange views and discuss various subjects.

To establish and sustain a stable mentoring partnership based on respect and mutual understanding, it is important to carefully match the mentoring couples on the basis of both functional and personal parameters. Hence, all candidates who apply for the programme (both mentees and mentors) fill in a detailed questionnaire that addresses not only the background (special courses and interests for pupils and parameters of study and scientific experience for mentors) and personal interests beyond school or university, but also and most importantly the expectations and wishes according to the future mentoring partner. So, the mentees can choose the study subject (or area, like natural sciences or engineering) of their mentor, his or her gender and other interests or characteristics, as far as possible in the framework of the pool of candidates.

Similarly, the mentors have an influence on the choice of their future mentee. The mentoring couples are matched by the programme coordinator on the basis of the answers given in the questionnaires and on the impressions from personal interviews with all candidates, to ensure a trustful and respectful partnership between mentee and mentor, founded on interests and personality of the participating persons. Since the mentors also serve as role models for the pupils, in addition to their function as an expert in a SET subject and personal advisor, they are chosen to fulfill some basic personal criteria and receive special training to comply with this role. As a preparation for their role as a mentor, the candidates accepted for mentoring take part in a two-day seminar, which consists of both the basics of mentoring with guidelines for establishing a fruitful mentoring partnership, and the conveyance of soft skills like communication and presentation techniques, conflict management and self reflection.

During the mentoring cycle, mentor and mentee maintain regular contact that can, due to the location of the mentee and the preferences of both partners, be established either by regular e-mails (at least once per week), internet chat (e.g. ICQ), phone calls, personal meetings or a combination of these options. So the participants are able to freely control the time invested into the mentoring relationship, while the regularity is assured by the minimal condition of a weekly contact. In these contacts, the partners exchange information and insight in their regular activities in study, research, or school, so that the mentees can "accompany" their mentor in her or his studies and get a closer connection to university, while the mentors see their subject and their own students career from a different perspective and get new insights. Although most of the mentoring conversation is focused on university and SET subjects, the participants are encouraged to speak also about any other fields of interest and activities.

Modules of TANDEMschool: MINT

Mentoring – The mentoring relation forms the main feature of the programme, providing the participants with a steady contact as a source of both advice and new impulses and possibility to exchange ideas and views.

Inspiration – Through insights into the research and science at the university as well as the participation in events in the framework of the programme, mentees are able to strengthen their interest in Mathematics, Engineering, Natural Science and Technology and discover opportunities on how to shape their own professional future in these fields. They also have access to a pool of additional information and activities offered by various institutions at the university.

Networking: Besides the contact with their personal mentor, the mentees are invited to regular network meetings, partly combined with special impulses like lectures or visits to research institutes. Thus, they also form contacts with people of similar interests and who are in comparable situations, far above the possibilities in the framework of their schools. Furthermore, there are also network meetings for the mentors as well as joint events, in which the mentees meet other mentors from different subjects, and have the chance to widen their view in addition to the mentoring itself. Thus, the mentees are early introduced to the benefits and importance of building up and maintaining personal networks as a factor for a successful career.

Training: The mentees take part in workshops on special areas of applied scientific and technical subjects lasting for 3 to 5 days, which provide a combination of theoretical seminars with practical hands on experiments, e.g. the summer and winter schools offered twice a year or the pupils' university at RWTH Aachen. Through the experimental work in small teams, networking among the pupils is further encouraged and strengthened.

Note that the module names, three of which are common for all mentoring programmes at RWTH Aachen University (see above) form the acronym MINT (German equivalent of SET).

Benefits for the mentees

As a direct result, the stimulation and strengthening of the interest in SET subjects can encourage pupils to choose scientific or technical subjects in school and become aware of possible applications in daily life.

The main focus, nevertheless, is guidance into a secure and suitable orientation regarding study subjects and future career prospects. Due to the practical information and experiences, the participants are empowered to choose a study subject according to their own interests, potentials and talents, and are encouraged to take the SET area into account. The activities in the programme provide them with a realistic image about the contents, opportunities and also the demands of studies in a specific SET subject, such that these insights help to facilitate the transition from school to university and to avoid wrong decisions. Thus, these realistic views combined with the presence of role models from the considered study fields can reduce the number of drop-out during the first years of studies (which is mostly caused by incomplete or wrong images of the study subject) and hence it can improve the success of the participants at university and also their satisfaction with the chosen subject.

Due to regular contact with an experienced person, the pupils enhance their personal and professional skills, especially communicating about technical contents and development of their personality by viewing their own situation from a different perspective outside school.

Through the activities of the programme such as meetings and workshops, the mentees are stimulated to build up a personal network and they obtain an early understanding of networking as an important factor for a career in science and technology.

The regular contact with a personal advising partner and role model widens the perspective of the mentees and brings them already in closer contact with university, so that they become familiar with the structures and opportunities of a technical university. These positive personal role models are of special importance for encouraging girls to consider a SET subject, since they can provide a good practice model for successful women in science and make them more visible to pupils. The participants get a deeper insight into research and science than other pupils and access and information about numerous other activities that the university offers to interested pupils and schools. Finally, they get to know RTWH Aachen University and the chosen field of study through "insiders".

Benefits for the mentors

It is important to note that the programme is not only designed for the profit and support of the mentees, but that the main targets are also benefits for mentors and their advancement in studies and science.

So the experience of mentoring helps the mentors to strengthen their consulting skills and the ability to pass on scientific knowledge to less experienced people, a capability that is especially crucial for activities in tutoring and teaching. A remarkable benefit for these and other soft skills also comes from the two-day seminar preparing the mentors for their task.

Through regular communication with the mentees, the mentors learn to see their own field of study from a different perspective and are newly motivated to reflect on their own career and decisions by sharing their own experiences with the mentee.

Due to the network meetings with other mentors and the joint meetings with the mentees, the mentors also enhance their personal networks and establish new contacts with people from other subjects and research areas, which can lead to fruitful inputs and new cooperations.

The mentors serve, at least to a certain extent, as personal role models for their mentees and so they have the possibility to actively support and promote young talent on their first steps towards their academic career, and get an insight into the current situation at schools, which might be of special interest for those involved in teaching and tutoring. Obviously, all these activities also contribute to a positive outside image of RWTH Aachen University and can increase its attractiveness for future students.

Curricular embedding

Besides a certificate for the soft skills obtained in the preparatory seminar, the mentors receive a second one for the mentoring itself. These certificates can be considered as valuable additional qualifications for future applications both inside and outside university, while certain mentors can also get their activities acknowledged and certified as qualified study performance, e.g. in credit points.

Since the programme is being embedded in the curricular courses of a growing number of subjects (usually in the framework of a module on soft skills or methodological competence), students from Bachelor and Master courses can obtain ECTS credit points (European Credit Transfer System) for their activities, while teaching students can have them credited as practical experiences in the compulsory curricular module "fascination in technology". Students from other courses (e.g. Diploma students) and doctoral students can use the certificates as additional qualifications.

This innovative approach to embed the mentoring activity into curricula activities creates an additional benefit for the mentors, increasing the attractiveness of the programme, so that the students consider mentoring, in the first place, not as an extra burden in their tight curriculum, but rather as an extra chance and possibility with variable time parameter.

As a part of the MINT cooperation concept within RWTH Aachen University's Institutional strategy, TANDEMschool is being funded in the framework of the German Excellence Initiative by the German Research Foundation.

Conclusion

A mentoring programme embedded in both the institutional strategy of the university and linked to further measures and incorporated in the curricular structure can serve as a good practice model for attracting more pupils to the SET area and reduce the drop-out rate, since it combines positive effects for pupils (future students) and the participating students, who are provided with important soft skills and crucial experiences.

The evaluation of the workshops in both summer and winter schools shows that these measures, combined with the mentoring, are especially effective to attract more girls to the SET area, offering practical experiences, insights and information into the subjects as well as positive role models from all stages of the academic career, thus also making female scientists and engineers visible to participants. Due to these realistic insights and personal experiences into the study opportunities and demands as well as research and daily work life in the chosen SET subjects, one can expect that these programmes not only can help to motivate more pupils to study SET subjects, but also may ensure sustainable effects in student retention and finally to help establish equal opportunities and gender justice within the scientific system.

References

acatech, VDI (ed.). 2009. Ergebnisbericht – Nachwuchsbarometer Technikwissenschaften. Munich/ Düsseldorf: acatech und VDI

Breuer, Elke & Leicht-Scholten, Carmen. 2011. Gender-Oriented Human Resources Development in International Cooperation – The Mentoring Programme TANDEMplusIDEA as a Model of Best Practice. In: Leicht-Scholten, Carmen; Breuer, Elke; Tulodetzki, Nathalie & Wolffram, Andrea (Hrsg.): Going Diverse: Innovative Answers to Future Challenges. Gender and Diversity Perspectives in Science, Technology and Business, Budrich UniPress Ltd.: Opladen: 195-208

Lämmerhirt, Marcel. 2009. Inspiring girls and boys in science, engineering and technology on virtual networks – The e-mentoring programme TANDEMkids at RWTH Aachen University, in this volume

Leicht-Scholten, Carmen. 2008. Exzellenz braucht Vielfalt – oder: wie Gender and Diversity in den Mainstream der Hochschulentwicklung kommt – Human Resources, Gender and Diversity Management an der RWTH Aachen. Journal Netzwerk Frauenforschung NRW 23: 33-39

Leicht-Scholten, Carmen et. al (ed.). 2008. Mentoring Handbuch – Ein Leitfaden In: http://www.igad.rwth-aachen.de/pdf/Mentoring-Handbuch_RWTH_Aachen.pdf [11.11.2009]

Leicht-Scholten, Carmen. 2011. Meeting Global Challenges – Gender and Diversity as Drivers for a Change of Scientific Culture. In: Leicht-Scholten, Carmen; Breuer, Elke; Tulodetzki, Nathalie & Wolffram, Andrea (Hrsg.): Going Diverse: Innovative Answers to Future Challenges. Gender and Diversity Perspectives in Science, Technology and Business, Budrich UniPress Ltd.: Opladen: 53-64

RWTH Aachen. 2008. Institutional Strategy: RWTH 2020: Meeting Global Challenges, not published, further information: http://www.exzellenz.rwth-aachen.de/aw/cms/home/Zielgruppen/~siy/ zukunftskonzept/?lang=en

RWTH Aachen. 2009a. Zukunftskonzept für die Lehre „Studierende im Fokus der Exzellenz. In: http://www.rwth-aachen.de/global/show_document.asp?id=aaaaaa aaaabycub [11.11.2009]

RWTH Aachen. 2009b. Zahlenspiegel 2008. 1st Edition. In: http://www.rwth-aachen.de/global/show_document.asp?id=aaaaaaaaaabqzxs [11.11.2009]

Statistisches Bundesamt. 2008. Bildung und Kultur – Nichtmonetäre hochschulstatistische Kennzahlen 1980-2007. Fachserie 11 Reihe 4.3.1. Statistisches Bundesamt: Wiesbaden

Statistisches Bundesamt. 2009. Bildung und Kultur – Studierende an Hochschulen – Wintersemester 2008/09. Fachserie 11 Reihe 4.1. Statistisches Bundesamt: Wiesbaden

How to change stereotypical images of science, engineering & technology? Results and conclusions from the european project MOTIVATION[66]

Felizitas Sagebiel (Germany), Carme Alemany (Spain), Jennifer Dahmen (Germany), Bulle Davidsson (Sweden), Anne-Sophie Godfroy-Genin (France), Gabriela Kol'veková (Slovakia), Cloé Pinault (France), Els Rommes (The Netherlands), Mariska Schönberger (The Netherlands), Anita Thaler (Austria), Natasa Urbancíková (Slowakia), Christine Wächter (Austria)

What images of disciplines and professions in science, engineering and technology (SET) are prevalent among adolescents and what factors influence them? These questions were addressed in the last two years by teams from Austria, France, Germany, Slovak Republic, Spain, Sweden and The Netherlands in the EC-funded project "MOTIVATION – promoting positive images of SET in young people under gender perspective".

Introduction

Research on SET and gender in media focussed on SET professionals so far. The new focus of MOTIVATION was to focus on general images of SET in youth magazines and television, in school and by the young people themselves taking initiatives of good practice on a national basis into account (Sagebiel 2008). Methods included document and media analysis, interviews, focus groups and drawings.

Youth magazines as SET learning fields

In a first phase of the project, in 2008, the consortium explored relevant youth magazines and analysed images of SET and gender quantitatively and qualitatively (Thaler 2009). In total 1.016 SET images of Austrian, French, German, Dutch and Slovak youth magazines were analysed. One remarkable insight is that technology plays a great role in youth magazines, but only

---

66 This article is reprinted by permission of responsible person. Smaller changes have been done by English language review by the publisher.

3.1% of the analysed images show SET as a job field; the rest represent SET products.

German magazine "BRAVO" (which is the most popular youth magazine in Germany and Austria) is partly overtly gender and SET stereotypical. For instance, vehicles are presented as male technology, showing males driving cars, motorbikes and even boats, females are mostly presented as co-drivers or even simply as models posing beside vehicles. "BRAVO GIRL!" has been identified as a magazine with a strong hetero-normative direction mainly aiming at girls and how they can appeal to boys (cf. Dahmen, Thaler 2009). Austrian magazine "Xpress" has less overt gender and technology stereotypes but in the analysis more subtle forms are visible (ibid.).

French magazine "Closer" presents SET as a job field only in training ads. In addition, SET is predominantly shown with males. Females in SET pictures are presented not only gender stereotypically but, moreover, in a sexist fashion. The second analysed French magazine 'Phosphore' presents SET equally with males and females. Dutch magazine "Girlz!" presents SET with competent females, but mostly competent with female connoted technologies. In contrast, the other Dutch magazine "Quest" presents SET mostly with males, and there especially male connoted technologies. In mixed gender groups or female representations in SET images, traditional gender roles are often reinforced.

Soap operas as informal SET education

In the second project phase, in 2009, the consortium analysed images of SET and gender in "soap operas" in all partner countries. Six of the seven analysed soap operas "Fisica o Quimica" (Spain), "Gute Zeiten? Schlechte Zeiten" (Germany), "Goede Tijden? Slechte Tijden" (Netherlands), "Panelák" (Slovakia), "Andra avenyn" (Sweden), and "Anna und die Liebe" (Austria) are offering similar results as the magazine analysis: Technology is often part of the stage set and seldom used in a meaningful way. The positive exception of our soap opera analysis is the French TV series "Plus Belle La Vie", which broaches the issue of SET in various ways, mostly via female and male SET professionals and up to date scientific and engineering stories (Thaler et al. 2009).

Gendered curricula and school books

In addition to interviews with high school students (age 13-18) and with teachers, school books for secondary schools were analysed. Books contain very few images of people (except in biology). There are sometimes some paragraphs about the history of science highlighting famous figures as Archimedes, Newton, Volta, etc. with very few female figures. Inside general education, biology is more related to real life than physics or chemistry. Maths appears as the "driest" discipline with almost no images, no context, and in some cases black and white books. When there are different curricula in general education, as in France, these trends are reinforced in the science curriculum, when the humanities curriculum has a very light scientific contents embedded in historical and social context.

The result is quite paradoxical: Dry and theoretical approaches are less attractive for pupils, but representations of science and scientists are not obviously gendered (except in the historical chapter). Contextualised approaches are more attractive, but tend to present gendered representations of science: Women are often the patient, the mother, the cashier, etc. when men are engineers, medicine doctors, etc. In almost all European countries, regulations or laws recommend that attention should be paid to gender issues in school books, however in reality it is not taken into account everywhere. Vocational education is very gendered and it is not so surprising to find very gendered representations in schoolbooks. The challenge would be to have an attractive presentation of sciences (including maths) in context, with an attention to gender and without abandoning the substance of the scientific contents. For vocational education, the issues would be a more gender-balanced representation of sciences; it would imply a gender neutral representation of professions.

Interest in and Image of SET

Besides the 77 interviews with pupils 10 focus group discussions were done to study how various youth cultures and cultural beliefs about how they perceive SET practicianors and students are potentially influencing pupils' choices about SET. 50 different kinds of pupils participated (boys, girls, interested in SET or not, heterosexual/non-heterosexual). Pupils in all countries seem to find SET sometimes interesting, e. g. as subject or hobby, but not as career to do "all the time" because it is "not fun", "not about people" and "monotonous". They consider SET jobs to be about "earning money", "getting opportunities" and it is "for/with men". These are for most pupils not the most relevant aspects of a future career. Opinions about SET do, however, to be somewhat depending on how much they like and how good their teachers are. In total almost all the boys in our sample indicated that they were interested in SET and only half of the girls were interested in SET.

Family relations seem to be mostly traditional in all countries. In all countries the father is most often by far considered the most competent SET person in the household with Slovakia as the biggest exception. It seems that pupils with a traditional work distinction in the family will make more traditional choices. Only a few of the pupils have the intention to work in a SET-profession, of whom most are boys (42%), versus 19% of the girls.

Pupils' images of SET persons are still very stereotypical, despite the increased numbers of women SET practitioner there are in society. Less than 4% of the drawings they were asked to do show a woman SET practitioner. This outcome conforms (or even reaches beyond) earlier "draw a scientist" (DAS) studies in other countries. Interestingly, of the girls who did draw a woman SET person, a relatively large number drew "their teacher" or "themselves".

Success of inclusion SET initiatives

Good practice examples to change images of SET professions among young people were identified for each partner country, founders and persons in charge of these initiatives were interviewed to learn more about their sustainability. Initiatives that have been running for many years have made a much stronger impression on young people than projects of shorter duration. This was true for selected German nation-wide established Girls' Day, a one-day activity for girls that offers insight into job fields where women are still underrepresented with the aim of weakening gender stereotypes in job decisions. Pupils who, because of inclusion initiatives, were keen on SET in the age group of 10 to 12 years', but who did not attend any further such initiative, were after some years not planning any SET education or profession. Initiatives including practical training seem to become more successful in encouraging pupils into SET than others.

Conclusions

Overall we can conclude that technology plays an important role in young people's lives. No wonder that those technological devices are part of media representations as well. Youth magazines and soap operas have lots of different possibilities to embed SET as meaningful topics. But only few producers use this chance, like in a job special section of the German youth magazine "BRAVO" or for an explosive storyline in the French soap opera "Plus Belle La Vie". Most youth media represent SET, and most often technology as well, in an accessory-style, like clothes or furniture they are used in the stage set of TV scenes or magazine pictures to represent modernity, where unfortunately the message too often is that possessing is more important then using and understanding (Thaler 2009). A lack of visible female SET role models also explains why pupils hardly ever draw a woman SET person. If they draw a woman engineer or scientist they indicate that this is a drawing of a specific person they know. More emphasis, thus, has to be put on showing careers and cultures of SET in youth media and to make women in SET and their professional life visible, not only to change the all too often still one-sided representations of SET as a male domain but also to bring SET on the adolescent's "radar of interesting job perspectives", for both boys and girls. For that reason, one aim of the project is to have an impact on youth media and to inform and motivate "persons in charge" of possibilities to show technology and SET professionals more often and also more appropriately. Our dissemination approach was bottom-up, thus leaflets, personal meetings, seminars were the best advertising of the objectives important to the motivation team. Non-academic journal or newspaper editors, even TV producers were interested in our research, and the project website provides further information: www.motivation-project.com.

References

Dahmen, Jennifer; Anita Thaler. 2009. Image is everything! Is image everything?! About perceived images of science, engineering and technology. In: Maartje van den Bogaard, Erik de Graf, Gillian Saunders-Smits (eds.): Proceedings of 37th Annual Conference of SEFI. "Attracting young people to engineering. Engineering is fun!" 1st-4th July 2009, Rotterdam. CD-Rom. ISBN 978-2-87352-001-4

Sagebiel, Felizitas. 2008. Motivation of young people for studying SET. The gender perspective. In: Proceedings of Sefi conference, Aalborg, 30.6.-3.7.2008 (CD) (Co-authors: Jennifer Dahmen, Bodil Davidsson, Anne-Sophie Godfroy-Genin, Els Rommes, Anita Thaler, Natasa Urbancíková)

Thaler, Anita. 2009. "Learning technology"? About the informal learning potential of youth magazines. In: Daniela Freitag, Bernhard Wieser, Günter Getzinger (eds.): Proceedings of the 8th Annual IAS-STS Conference on Critical Issues in Science and Technology Studies. Graz: DRom. ISBN: 978-3-9502678-1-5. Download: http://www.ifz.tugraz.at/index_en. php/article/articleview/1817/1/58 [28.10.2009]

Thaler, Anita; Dahmen, Jennifer & Pinault, Cloé. 2009. European media images of science, engineering and technology. IFZ – Electronic Working Papers 2-2009. Download: http://www.ifz.tugraz.at/index.php/article/articleview/1621/1/154) [16.11.2009]

Authors' short biographies

Dr. Frank S. Becker was born in 1952 in Marburg, Germany. After studying physics at the University of Karlsruhe, he worked at the Max-Planck Society for the Promotion of Science in Munich. In 1981, he received a PhD from the University of Munich and began working in the Central R&D Department of Siemens AG. After a number of different assignments, he moved in March 2003 to the Corporate Personnel Department, where he became spokesman for university-related topics like the European Higher Education Area and the industry requirements regarding engineering education. He represented Siemens in a number of German professional organizations (ZVEI, VDI, ASIIN, BDA). From July 2005 until November 2012, he has been responsible at Siemens' Corporate Communications and Government Affairs for topics related to higher education. (Chairman of the division "Engineering Education" of the Association of German Engineers (VDI), Verein Deutscher Ingenieure e.V., VDI-Platz 1, 40468 Düsseldorf, Germany, franksbecker@gmail.com)

Flora Di Martino, after the degree in Geology (1991) and a collaboration with the Department of Geophysics and Volcanology at Università degli Studi di Napoli Federico II, in 1992 Flora started to work at Fondazione IDIS-Città della Scienza in Naples. As project manager she is involved in EU funded projects (Science in Society). She is currently coordinator of the Education Sector and works for a better use of Information & Communication Technologies in schools. (dmartino@cittadellascienza.it)

Judith Ebach is a psychologist and consultant at the Research Department at the University of Bonn. Since her diploma (1987) she worked in several organisations which focus on female career decisions. Additional key activities during her professional career are the development and evaluation of diagnostic instruments especially the assessment of systemic thinking and problem-solving competencies in school related environments, e.g. in physic lessons. From 2002 to 2009 she led the coordination office of the Ada-Lovelace-Project, a measure which intends to motivate girls and young women to choose a career in science, engineering or mathematics. (j.ebach@uni-bonn.de)

PD Dr. Martina Endepohls-Ulpe is senior lecturer at the University of Koblenz-Landau, Campus Koblenz, Institute for Psychology. She teaches educational and developmental psychology to teacher students and students of pedagogy. In the last years she published several articles on the topic of technology and science education for girls. Other topics of research and publishing have been gender differences in academic achievement, giftedness, gifted education and consequences of divorce for parents and children from which she presented results on international conferences regularly. She acts as scientific advisor for the Ada-Lovelace-Project, a measure which intends to motivate girls and young women to choose a career in science, engineering or mathematics, since 2005. (endepohl@uni-koblenz.de)

Prof. Dr. Anne-Sophie Godfroy, after studying philosophy at the Ecole Normale Superieure de Paris and completing a PhD in philosophy at University Paris-Sorbonne, Dr. Anne-Sophie Godfroy is associate professor at University Paris-Est-Creteil where she teaches at the pedagogical Institute (IUFM). She is researcher in the new research centre "Sciences, Norms, Decision" (CNRS and Paris Sorbonne). Her main research interests are methodology for international comparisons, science in society, gender and science education. She has participated in European research projects for ten years. Her publications focus methodological as well as gender and engineering questions. She edited the very important proceedings of PROMETEA final conference "Women in Engineering and Technology Research". (Ecole Normale Supérieure de Cachan & University Paris-Est-Creteil (Paris, France)). (anne-sophie.godfroy@u-pec.fr)

Daniele Gouthier, after a Phd in mathematics at SISSA (International School for Advances Studies, Trieste, Italy), and since 1996, Daniele works as science communicator. Researching on the communication of science for the young public, he took part of national and international projects and initiatives (Scienza Under18, GAPP, Sedec). He's editor in chief of Scienza Express (www.scienzaexpress.it).

Dr. Kathrin Gräßle, PhD in educational science, studies in Tübingen and Kopenhagen, Diploma in pedagogy with speciality of education in Dortmund, thesis in adult education/ educational consulting at the University of Duisburg-Essen, besides occupational employment. After several years of adult education she was employed as study advisor at the University of Duisburg-Essen leader of the project "Summer University for Women in Science and Technology". She is employed in the Ministry for Innovation, Science, Research and Technology of Nothrhine-Westfalia and engaged in the unit "Research and Education, Transfer from School to High School, Future through Innovation". (kathrin.graessle@web.de)

Prof. Dr. Albert Gras-Martí, emeritus professor of Applied Physics at the University of Alacant, with a PhD in Physics in the field of Atomic Collisions in Solids. Recent research interests are in Physics Education Research, especially in the applications of Information and Communication Technologies to Science Education and eLearning in general, both in face to face and distant education. Present activities involve consulting in educational projects as well as participating in projects of science and mathematics education for advanced high-school students in non-traditional environments. (SAEF-CI, http://saef-ci.com)

Dr. Àgueda Gras-Velazquez is the Science Programme Manager of European Schoolnet (EUN) in charge of overseeing and coordinating all the

Maths and Science projects run by EUN. Additionally, at the moment, she is in charge of the day to day management of Scientix (DG Research's community for science education in Europe) the school piloting of the FP7 project inGenious (ECB) and EUN's involvement in other EC funded projects like Pathway, Nanopinion, Go-Lab and Global Excursion to name a few. Prior to joining EUN in May 2008, she worked for over five years as an independent eLearning Professional, as Tutor, Content designer, IT manager, Administrator, Project Manager and Consultant for international projects. She has co-authored several papers in the area of Science Education Research, particularly on teaching applications of ICT, Fraud in Science and Scientific paradigm shifts in medicine and women. She has a PhD in Astrophysics from Trinity College Dublin, which she carried out at the Dublin Institute for Advanced Studies. (European Schoolnet, http://europeanschoolnet.org (Brussels, Belgium)). (agueda.gras@eun.org)

Karin Griffiths (nee Grabarz) is currently holding a position as a research assistant in the BMBF funded project 'Gender Disparities in the Occupational Career of Mathematicians and Physicists' at the University of Bielefeld. She studied sociology and psychology at the RWTH Aachen, University of Trier, University of Gdansk (Poland) and graduated in sociology in 2011 at the University of Bielefeld. Her major fields of interests cover empirical research on inequalities focusing on work and higher education, characteristics and developments of occupational careers as well as women and girls in STEM fields. (IFF, University of Bielefeld (Germany)). (kgriffiths@uni-bielefeld.de)

Gehrt Hartjen, Dipl.-Math., he studied mathematics and physics at RWTH Aachen University (Diplom theses in computer algebra on "Variational calculus and conservation laws with MAPLE"(2001)). From 2002 to 2007, he was a scientific assistant at the Chair B for Mathematics, RWTH Aachen University. Since 2008, he works at RWTH Aachen University's *Integration Team – Human Resources, Gender and Diversity Management*, where he has designed and coordinated the mentoring programme TANDEMschool and further activities for pupils in the SET (MINT) area. RWTH Aachen. (gehrt.hartjen@rwth-aachen.de)

Alexa Joyce is a specialist in ICT in education and currently working as Senior Business Development and Communications Manager at EUN. Previously she worked at UNESCO Bangkok, as project coordinator of the pilot Southeast Asian Schoolnet, focusing on ICT-based international cooperation for learning science, maths and English. During her time in Bangkok, Alexa was also involved as a trainer and consultant in UNESCO's Next Generation of Teachers and Teacher Training projects, supporting teachers in transforming their pedagogies to a more ICT-based, student-centred approach, and co-organised the Innovative Teachers conference with Microsoft Singapore. She has also consulted for UNESCO International Institute of Educational Planning, Paris, and for the OECD Centre for Educational Research and Innovation. She has also worked for EUN for 9 years, managing numerous EC-funded projects and multi-stakeholder partnerships. She has a Masters in Biological Sciences from the University of Oxford and an MBA from Solvay Business School, Brussels. (alexa.joyce@eun.org)

Prof. Dr. Ursula Kessels studied Psychology. After her graduation in 1998 she worked on several research projects, funded by the Deutsche Forschungsgemeinschaft, at different universities. In 2001 she took her PhD in Psychology at the Freie Universität Berlin with a doctoral dissertation on single-sex classes in physics. In 2007 she was habilitated at the Freie Universität Berlin for Psychology. From 2009 to 2013 she was professor for Educational Psychology at the University of Cologne. Since 2013 she is professor for Educational Research at the Freie Universität Berlin. Her main research interests concern the interplay of (gender-) identity and academic achievement/ interests. (ursula.kessels@uni-koeln.de)

Dr. Gabriela Koľveková, Ing., PhD has studied economics at Technical University of Košice, Faculty of Economics (MA in Finance, banking and investment). She obtained her Ph.D. degree at Czech University of Life Sciences Prague, Faculty of Economics and Management (specialization: Regional and Social Development). She has published over 70 papers or chapters in books, journals or proceedings. She attended several study stays e.g. in Malta, France, Poland. Her interest in regional economy was deepened by the various research projects, focussing on a great variety of concrete issues: Economic, ecology and social problems, regional development, public private partnership, start-up entrepreneurs, human capital and labour market, and gender studies in engineering. (Technical University of Košice, Faculty of Economics, Němcovej 32, 040 01 Košice (Košice, Slovak Republic). (Gabriela.Kolvekova@tuke.sk)

Dr. Sybille Krummacher is an experimental physicist at Research Centre Jülich, Germany (FZJ). She has been active in the pioneering times of several synchrotron radiation laboratories in France, the US and Germany, and was also involved in the planning of the European Synchrotron Radiation Facility, ESRF, now in Grenoble. Her field of interest has evolved from electron spectroscopy of free atoms and molecules via surfaces and thin films to that of atomic clusters, including fullerenes. In 2011 the French Minister of Education nominated her "Knight in the Order of the Academic Palms" for her extraordinary commitment for French-German cooperation and understanding. She has been responsible for the bilateral cooperation in science and technology between Germany and countries of the Middle East. In addition to more than 20 years of basic research in international and interdisciplinary surroundings she has also been active in the promotion of women in science, with a special focus on women in publicly funded research centres. (s.krummacher@fz-juelich.de)

Dr. Bettina Langfeldt is senior researcher at the Helmut Schmidt University – University of the Federal Armed Forces Hamburg (Germany) at the Faculty of Humanities and Social Sciences. She studied at the Philipps-University in Marburg (Germany), afterwards she worked at GESIS – Leibniz-Institute for the Social Sciences in Mannheim (Germany) and at the Justus-Liebig-University in Gießen (Germany), where she got her PhD in 2006. As an expert in the field of research methods and statistics she advised, conducted or led a great number of research projects. Besides methodological topics her interests in applied research range from the sociology of work to organizational problems in higher education systems and gender aspects in science, technology, engineering, and mathematics (STEM). (bela@hsu-hh.de)

Marcel Lämmerhirt is a scientific employee at Aachen RWTH University since 2008. He studied economic geography, sociology and economics as well as technical communication at RWTH Aachen University. In the time from 2009 to 2011 he was the coordinator of the mentoring programme TANDEMkids at the executive department "Integration Team for Gender and Diversity Management" and was therewith responsible for the concept and the development of the programme. Since 2011 he is occupied at the Center for Computing and Communication as technical sub-project manager in the university-wide project "PuL" which aims at the improvement and the IT support of organizational processes in higher education. (marcel.laemmerhirt@rwth-aachen.de)

Prof. Dr. Carmen Leicht-Scholten is professor for "Gender and Diversity in Engineering" at the Faculty of Civil Engineering at RWTH Aachen University. Since May 2012, Prof. Leicht-Scholten is Dean of Study Affairs of the Faculty of Civil Engineering. She is member of numerous national and international advisory boards, for example, the strategy council "Diversity" of the Austrian Ministry for Science and Research, at the EU as well as the "Women's Research Network NRW". From 2010 until 2011 Prof. Leicht-Scholten held a visiting professorship in "Gender and Diversity Management in Engineering" at the Technical University Berlin. From 2007 until 2010 she was head of the scientific unit "Integration Team, Human Resources, Gender and Diversity Management" (IGaD) at RWTH Aachen University. Her research focuses are gender and diversity in scientific and technical research, gender and diversity in (scientific) organisations as well as gender and diversity in engineering. (carmen.leicht@gdi.rwth-aachen.de)

Federica Manzoli got her degree in Communication Science in 1998 (University of Siena, Italy). While working in the marketing research field, in 2003 she took the Master in Science Communication at Sissa (International School for Advanced Studies, Italy); there she started her teaching activity on the public perception and communication of S&T. In 2011 she discussed her PhD in Science-and-Society at the University of Milan on the topic of the public perception and communication of climate change. Since 2010 she teaches at the Master in Science Communication at University Milano Bicocca. Since her first degree, she collaborates in SiS European projects (FP6-7). (federica.manzoli@gmail.com)

Ilse Marschalek is a sociologist with several years of experience in international studies in the 5th, 6th and 7th FP of the EC. Additionally she initiated and coordinated many national studies in the fields of participation, gender and migration. At Centre for Social Innovation (ZSI) she is project coordinator at the technology and knowledge department, carrying out a range of projects at the interface between technological and societal innovations. Currently she is involved in the science communication study NANOPINION (communication of nanotechnology) and the virtual science infrastructure in the GLOBALexcursion project. Research fields: Inter- and transdisciplinarity, participatory action research and involvement of non scientific persons into research processes; science communication, technology enhanced learning, participatory evaluation and research processes. (marschalek@zsi.at)

Dr. Anina Mischau holds a PhD in sociology. From 2002-2012 she worked as a senior researcher at the Interdisciplinary Center of Female Research and Gender Studies (IFF) at the University of Bielefeld. Currently she is a visiting professor for "Gender Studies in Mathematics and Didactics of Mathematics" at the Department of Mathematics and Computer Science at FU Berlin. Her main research and teaching areas are gender disparity in higher education and science, gender in mathematics, natural sciences and technology, gender competence for teacher training in mathematics and equal opportunity policies in higher education and science. (Freie Universität Berlin (Berlin, Germany)). (amischau@mi.fu-berlin.de)

Petra Moser, born in 1980, is research associate at the Centre for Social Innovation (ZSI) in the unit *'work and equal opportunities'* in Vienna. She studied Sociology and Communication Science at the University of Vienna and received her Master degree (Magistra) in 2008 after field research on rural livelihood transition and changing policies in Lao PDR. She is currently engaged in projects in the field of demographic change, social economy, lifelong learning and digital media. She has experiences in evaluation of European projects and project management. Furthermore she has been involved in the European funded NanoYou project (7th framework programme) focussing on young people and science communication using the example of nanotechnology. (moser@zsi.at)

Prof. Dr. Sylvia Neuhäuser-Metternich, professor of social competencies, mentoring and gender mainstreaming at the Dortmund University of Applied Sciences and Arts, Department of Information Technology and Electrical Engineering from 2003 until 2011; educated and trained in psychology at the University of Mainz, Ph.D. at the University of Giessen; chair of Ada Lovelace Mentoring-Association, an organisation dedicated at motivating young women to choose a career in SEM as well as building a network and strategic alliances with women and men from 2001 until 2011; author of books about communication in the workforce. (neumett@fh-dortmund.de)

Donato Ramani has a degree in Biological Sciences with an emphasis in biochemistry. He obtained a Master Degree in Science Communication in 2006. He is the project manager of SISSA's Master in Scientific Digital Journalism and collaborates with Italian magazines as freelance journalist. (donato.ramani@gmail.com)

Barbara Roth is head of the department of social education/gender justice/prevention at the Pedagogical Institute City of Munich. Her main skills are in transferring current topics of public interest, like gender equality, from scholarly pieces into teaching in schools in Munich. She studied Architecture, Construction Engineering, Pedagogics, Psychology and Sports, finished with a so called "Staatsexamen HLB" a state exam to become a teacher at special vocational schools in Bavaria. Postgraduate courses: Diploma in Public Relation (BAW), Gender Studies, business administration and law. She worked as a director for a not-for-profit-organisation, teacher at vocational schools, director of an adventure and experimental learning training centre. (barbara.roth@muenchen.de)

Prof. Dr. Felizitas Sagebiel, PhD in sociology, is social scientist and associate professor in the Faculty of Educational and Social Science (University of Wuppertal). Since 2001 responsible roles in several European Commission Projects from the 5th, the 6th to the 7th Framework Programme, INDECS, Womeng, PROMETEA, MOTIVATION (coordinator), Tender "Meta-analysis of gender and science research" (www.genderandscience. org/web/index.php) (report on "gender stereotypes"); research and publication on gender in science and engineering (education and professional sphere), msculinities, men's networks. Member of the board of Network Women and Gender Studies Northrhine-Westfalia (Germany), responsible for the board of the publisher LIT of the series "Gender interdisciplinary" (Faculty of Educational and Social Science, Bergische Universität Wuppertal (Wuppertal, Germany)). (sagebiel@uni-wuppertal.de)

Prof. Dr. Minna Salminen-Karlsson, PhD in Education, associate professor in Sociology and currently working at the Centre for Gender Research at Uppsala University, has researched gender in engineering education and engineering workplaces, as well as gender in other kinds of technical education, such as technology centres for schoolchildren and computer courses in adult education. She has studied reform initiatives in engineering education and lectured widely on gender issues in various institutes of technology. She has also studied gender issues in high tech workplaces both in private business and in academia, in Sweden and abroad. Her current research deals with gender issues in Swedish-Indian cooperation in ICT industry. She also coordinates an FP7 project, FESTA, aimed at improving women's situation in academia. (Minna.Salminen@gender.uu.se)

Magdalena Strasser, born in 1984 in Upper Austria, has been research assistant at the Centre for Social Innovation (ZSI) in the unit '*technology and knowledge*' since October 2008. She studied Psychology at the University of Vienna and graduated in 2009. From 2009 -2011 she passed the postgraduate programme "Clinical and Health Psychology" at the University of Vienna. (Zentrum für Soziale Innovation (Vienna, Austria)). (strasser@zsi.at)

Prof. Dr. Nataša Urbančíková, M.Sc., PhD is an Associate Professor at the Faculty of Economics (Head of Department of Regional Science and Management). She is Professor in the area of Business Administration and Management. She studied Microelectronics at the Technical University of Kosice and Business Administration and Management. Her research fields are in the human and social capital, HRD, regional science, continuing education (including open and distance learning), marketing and microelectronics (published more than 100 papers in these fields). She has extended experience in gender and science. She was a member of research team of FP5,6, and 7 projects related to gender and science. She was a national expert for "Meta-analysis of gender and science research" (RTD-PP-L4-2007-1) project commissioned by the European Commission. Technical University of Košice, Faculty of Economics, Němcovej 32, 040 01 Košice (Košice, Slovak Republic) (Natasa.Urbancikova@tuke.sk)

Prof. Dr. Susana Vazquez-Cupeiro, PhD in Sociology (University of London) and MPhil in Gender Studies (Trinity College Dublin). Currently she is professor in the Faculty of Education at the Universidad Complutense of Madrid (UCM). She is member of the Research Groups, "Employment, Gender and Social Cohesion" (EGECO) and "Sociological Analysis of Education" (ASE), and the Complutense Institute of Sociology to the Study of Social Transformations (TRANSOC). She has participated in research projects related to the system of gender and science, academic and scientific trajectories, the gender digital divide and educational inequalities. Recent Publications with other authors: "Praxis feminista on line contra la violencia de género en España. Una práctica política efectiva de agencia femenina en la Red", in TELOS (Journal of Communication and Innovation) n. 92 (2012) and Meta-analysis of gender and science research. Directorate-General for Research and Innovation, European Commission: Brussels (2012) (susana.vazquezcup@gmail.com).